Sex and Repression in Savage Society

'Malinowski altered the whole mode and purpose of ethnographic enquiry.'

Edmund Leach

'The present essay attempts to put Freud's theories to the test by examining them in the light of the mental habits of the harmless Trobrianders. . . . Several years' contact with Melanesians, backed by the power to converse with them freely, gives Malinowski the best right to be heard as a reporter of facts which, it must be admitted, would escape nine trained observers out of every ten.'

The Times Literary Supplement

'This work is a most important contribution to anthropology and psychology, and it will be long before our textbooks are brought up to the standard which is henceforth indispensable.'

Saturday Review

Bronislaw
Malinowski

Sex and Repression in
Savage Society

 London and New York

First published 1927
by Kegan Paul, Trench, Trübner & Co. Ltd
First paperback edition published 1960
by Routledge and Kegan Paul

First published in Routledge Classics 2001
by Routledge
2 Park Square, Milton Park, Abingdon, Oxon OX14 4RN
270 Madison Ave, New York, NY 10016

Reprinted 2007,2008

*Routledge is an imprint of the Taylor & Francis Group,
an informa business*

Typeset in Joanna by RefineCatch Limited, Bungay, Suffolk
Printed and bound in Great Britain by
TJ International Ltd, Padstow, Cornwall

British Library Cataloguing in Publication Data
A catalogue record for this book is available from the British Library

Library of Congress Cataloging in Publication Data
A catalog record for this book has been applied for

ISBN 10: 0-415-25554-6
ISBN 13: 978-0-415-25554-7

TO
MY FRIEND
PAUL KHUNER
New Guinea, 1914, *Australia*, 1918,
South Tyrol, 1922.

CONTENTS

PREFACE

The doctrine of psycho-analysis has had within the last ten years a truly meteoric rise in popular favour. It has exercised a growing influence over contemporary literature, science, and art. It has in fact been for some time the popular craze of the day. By this many fools have been deeply impressed and many pedants shocked and put off. The present writer belongs evidently to the first category, for he was for a time unduly influenced by the theories of Freud and Rivers, Jung, and Jones. But pedantry will remain the master passion in the student, and subsequent reflection soon chilled the initial enthusiasms.

This process with all its ramifications can be followed by the careful reader in this little volume. I do not want, however, to raise expectations of a dramatic *volte-face*. I have never been in any sense a follower of psycho-analytic practice, or an adherent of psycho-analytic theory; and now, while impatient of the exorbitant claims of psycho-analysis, of its chaotic arguments and tangled terminology, I must yet acknowledge a deep sense of indebtedness to it for stimulation as well as for valuable instruction in some aspects of human psychology.

Psycho-analysis has plunged us into the midst of a dynamic theory of the mind, it has given to the study of mental processes a concrete turn, it has led us to concentrate on child psychology and the history of the individual. Last but not least, it has forced upon us the consideration of the unofficial and unacknowledged sides of human life.

The open treatment of sex and of various shameful meanesses and vanities in man—the very things for which psycho-analysis is most hated and reviled—is in my opinion of the greatest value to science, and should endear psycho-analysis, above all to the student of man; that is, if he wants to study his subject without irrelevant trappings and even without the fig leaf. As a pupil and follower of Havelock Ellis, I for one shall not accuse Freud of 'pan-sexualism'—however profoundly I disagree with his treatment of the sex impulse. Nor shall I accept his views under protest, righteously washing my hands of the dirt with which they are covered. Man is an animal, and, as such, at times unclean, and the honest anthropologist has to face this fact. The student's grievance against psycho-analysis is not that it has treated sex openly and with due emphasis, but that it has treated it incorrectly.

As to the chequered history of the present volume, the first two parts were written much earlier than the rest. Many ideas laid down there were formed while I was engaged in studying the life of Melanesian communities on a coral archipelago. The instructions sent to me by my friend Professor C. G. Seligman, and some literature with which he kindly supplied me, stimulated me to reflect on the manner in which the Œdipus complex and other manifestations of the 'unconscious' might appear in a community founded on mother-right. The actual observations on the matrilineal complex among Melanesians are to my knowledge the first application of psycho-analytic theory to the study of savage life, and as such may be of some interest to the student of man, of his mind and of his culture. My conclusions are couched in a terminology more psycho-analytic than I should like to use now. Even so I do not go much beyond such words as 'complex' and 'repression', using both in a perfectly definite and empirical sense.

As my reading advanced, I found myself less and less inclined to accept in a wholesale manner the conclusions of Freud, still less those of every brand and sub-brand of psycho-analysis. As an anthropologist I feel more especially that ambitious theories with regard to savages, hypotheses of the origin of human institutions and accounts of the history of culture, should be based on a sound knowledge of primitive life, as well as of the unconscious or conscious aspects of the human mind. After all neither group-marriage nor tetemism, neither avoidance of mother-in-law nor magic happen in the 'unconscious'; they are all solid sociological and cultural facts, and to deal with them theoretically requires a type of experience which cannot be acquired in the consulting room. That my misgivings are justified I have been able to convince myself by a careful scrutiny of Freud's *Totem and Taboo*, of his *Group-Psychology and the Analysis of the Ego*, of *Australian Totemism* by Róheim and of the anthropological works of Reik, Rank, and Jones. My conclusions the reader will find substantiated in the third part of the present book.

In the last part of the book I have tried to set forth my positive views on the origins of culture. I have there given an outline of the changes which the animal nature of the human species must have undergone under the anomalous conditions imposed upon it by culture. More especially have I attempted to show that repressions of sexual instinct and some sort of 'complex' must have arisen as a mental by-product of the creation of culture.

The last part of the book, on Instinct and Culture, is in my opinion the most important and at the same time the most debatable. From the anthropological point of view at least, it is a

xii PREFACE

pioneering piece of work; an attempt at an exploration of the
'no-specialist's-land' between the science of man and that of
the animal. No doubt most of my arguments will have to be
recast, but I believe that they raise important issues which will
sooner or later have to be considered by the biologist and animal
psychologist, as well as by the student of culture.

As regards information from animal psychology and biology I
have had to rely on general reading. I have used mainly the works
of Darwin and Havelock Ellis; Professors Lloyd Morgan, Herrick,
and Thorndike; of Dr. Heape, Dr. Köhler and Mr. Pyecroft, and
such information as can be found in the sociological books of
Westermarck, Hobhouse, Espinas and others. I have not given
detailed references in the text and I wish here to express my
indebtedness to these works; most of all to those of Professor
Lloyd Morgan, whose conception of instinct seems to me the
most adequate and whose observations I have found most useful.
I discovered too late that there is some discrepancy between my
use of the terms instinct and habit and that of Professor Lloyd
Morgan, and in our respective conceptions of plasticity of instincts. I
do not think that this implies any serious divergence of opinion.
I believe also that culture introduces a new dimension in the
plasticity of instincts and that here the animal psychologist can
profit from becoming acquainted with the anthropologist's
contributions to the problem.

I have received in the preparation of this book much stimula-
tion and help in talking the matter over with my friends Mrs.
Brenda Z. Seligman of Oxford; Dr. R. H. Lowie and Professor
Kroeber of California University; Mr. Firth of New Zealand; Dr.
W. A. White of Washington, and Dr. H. S. Sullivan of Baltimore;
Professor Herrick of Chicago University, and Dr. Ginsberg of the
London School of Economics; Dr. G. V. Hamilton and Dr. S. E.
Jelliffe of New York; Dr. E. Miller of Harley Street; Mr. and Mrs.
Jaime de Angulo of Berkeley, California, and Mr. C. K. Ogden of
Cambridge; Professor Radcliffe-Brown of Cape Town and

Sydney, and Mr. Lawrence K. Frank of New York City. The field-work on which the book is based has been made possible by the munificence of Mr. Robert Mond.

My friend Mr. Paul Khuner of Vienna, to whom this book is dedicated, has helped me greatly by his competent criticism which cleared my ideas on the present subject as on many others.

B. M.

Department of Anthropology,

London School of Economics,

February, 1927.

'After ignoring impulses for a long time in behalf of sensations, modern psychology now tends to start out with an inventory and description of instinctive activities. This is an undoubted improvement. But when it tries to explain complicated events in personal and social life by direct reference to these nature powers the explanation becomes hazy and forced . . .

'We need to know about the social conditions which have educated original activities into definite and significant dispositions before we can discuss the psychological element in society. This is the true meaning of social psychology . . . Native human nature supplies the raw materials but custom furnishes the machinery and the designs . . . Man is a creature of habit, not of reason nor yet of instinct.

'The treatment of sex by psycho-analysts is most instructive, for it flagrantly exhibits both the consequences of artificial simplification and the transformation of social results into psychic causes. Writers, usually male, hold forth on the psychology of woman as if they were dealing with a Platonic universal entity . . . They treat phenomena, which are peculiarly symptoms of the civilization of the West at the present time, as if they were the necessary efforts of fixed native impulses of human nature.'

JOHN DEWEY, in *Human Nature and Conduct*

Part I

The Formation of a Complex

1

THE PROBLEM

Psycho-analysis was born from medical practice, and its theories are mainly psychological, but it stands in close relation to two other branches of learning—biology and the science of society. It is perhaps one of its chief merits that it forges another link between these three divisions of the science of man. The psychological views of Freud—his theories of conflict, repression, the unconscious, the formation of complexes—form the best elaborated part of psycho-analysis, and they cover its proper field. The biological doctrine—the treatment of sexuality and of its relation to other instincts, the concept of the 'libido' and its various transformations—is a part of the theory which is much less finished, less free from contradictions and lacunæ, and which receives more criticism, partly spurious and partly justified. The sociological aspect, which most interests us here, will deserve more attention. Curiously enough, though sociology and anthropology have contributed most evidence in favour of psycho-analysis, and though the doctrine of the Œdipus complex has obviously a sociological aspect, this aspect has received the least attention.

Psycho-analytic doctrine is essentially a theory of the influence of family life on the human mind. We are shown how the passions, stresses and conflicts of the child in relation to its father, mother, brother and sister result in the formation of certain permanent mental attitudes or sentiments towards them, sentiments which, partly living in memory, partly embedded in the unconscious, influence the later life of the individual in his relations to society. I am using the word *sentiment* in the technical sense given to it by Mr. A. F. Shand, with all the important implications which it has received in his theory of emotions and instincts.

The sociological nature of this doctrine is obvious—the whole Freudian drama is played out within a definite type of social organization, in the narrow circle of the family, composed of father, mother, and children. Thus the *family complex*, the most important psychological fact according to Freud, is due to the action of a certain type of social grouping upon the human mind. Again, the mental imprint received by every individual in youth exercises further social influences, in that it predisposes him to the formation of certain ties, and moulds his receptive dispositions and his creative power in the domains of tradition, art, thought, and religion.

Thus the sociologist feels that to the psychological treatment of the complex there should be added two sociological chapters: an introduction with an account of the sociological nature of family influences, and an epilogue containing the analysis of the consequences of the complex for society. Two problems therefore emerge for the sociologist.

First problem. If family life is so fateful for human mentality, its character deserves more attention. For the fact is that the *family* is not the same in all human societies. Its constitution varies greatly with the level of development and with the character of the civilization of the people, and it is not the same in the different strata of the same society. According to theories current even today among anthropologists, the family has changed enormously during the development of humanity, passing from its first promiscuous form, based on sexual and economic communism, through 'group-family' based on 'group-marriage', 'consanguineous family', based on 'Punalua marriage', through the *Grossfamilie* and clan kindred to its final form in our present-day society—the individual family based on monogamous marriage and the *patria potestas*. But apart from such anthropological constructions which combine some fact with much hypothesis, there is no doubt that from actual observation among present-day savages we can see great variations in the constitution of the family. There are differences depending on the distribution of *power* which, vested in a varying degree in the father, give the several forms of patriarchy, or vested in the mother, the various sub-divisions of mother-right. There are considerable divergencies in the methods of counting and regarding *descent*—matriliny based on ignorance of fatherhood and patriliny in spite of this ignorance; patriliny due to power, and patriliny due to economic reasons. Moreover, differences in settlement, housing, sources of food supply, division of labour and so on, alter greatly the constitution of the human family among the various races and peoples of mankind.

The problem therefore emerges: do the conflicts, passions and attachments within the family vary with its constitution or do they remain the same throughout humanity? If they vary, as in fact they do, then the nuclear complex of the family cannot remain constant in all human races and peoples; it must vary with the constitution of the family. The main task of psycho-analytic theory is, therefore, to study the limits of the variation; to frame the appropriate formula; and finally, to discuss the outstanding types of family constitution and to state the corresponding forms of the nuclear complex.

With perhaps one exception,[1] this problem has not yet been raised, at least not in an explicit and direct manner. The complex exclusively known to the Freudian School, and assumed by them to be universal, I mean the Œdipus complex, corresponds essentially to our patrilineal Aryan family with the developed *patria potestas*, buttressed by Roman law and Christian morals, and accentuated by the modern economic conditions of the well-to-do bourgeoisie. Yet this complex is assumed to exist in every savage or barbarous society. This certainly cannot be correct, and a detailed discussion of the first problem will show us how far this assumption is untrue.

The second problem. What is the nature of the influence of the family complex on the formation of myth, legend, and fairy tale, on certain types of savage and barbarous customs, forms of social organization and achievements of material culture? This problem has been clearly recognized by the psycho-analytic writers who have been applying their principles to the study of myth, religion, and culture. But the theory of how the constitution of the family influences culture and society through the forces of the family complex has not been worked out correctly. Most of the views bearing on this second problem need a thorough revision from the sociological point of view. The concrete solutions, on the other hand, offered by Freud, Rank and Jones of the actual mythological problems are much sounder than their general principle that the 'myth is the secular dream of the race'.

Psycho-analysis, by emphasizing that the interest of primitive man is centred in himself and in the people around him, and is of a concrete and dynamic nature, has given the right foundation

---

[1] I refer to Mr. J. C. Flügel's *The Psycho-Analytic Study of the Family*, which, though written by a psychologist, is throughout orientated in the sociological direction. The later chapters, especially XV and XVII, contain much which is an approach to the present problem, although the writer does not formulate it explicitly.

to primitive psychology, hitherto frequently immeshed in a false view of the dispassionate interest of man in nature and of his concern with philosophic speculations about his destiny. But by ignoring the first problem, and by making the tacit assumption that the Œdipus complex exists in all types of society, certain errors have crept into the anthropological work of psycho-analysts. Thus they cannot reach correct results when they try to trace the Œdipus complex, essentially patriarchal in character, in a matrilineal society; or when they play about with the hypotheses of group-marriage or promiscuity, as if no special precautions were necessary when approaching conditions so entirely foreign to the constitution of our own form of family as it is known to psycho-analytic practice. Involved in such contra-dictions, the anthropologizing psycho-analyst makes a hypo-thetical assumption about some type of primitive horde, or about a prehistoric prototype of the totemic sacrifice, or about the dream character of the myth, usually quite incompatible with the fundamental principles of psycho-analysis itself.

Part I of the present work is essentially an attempt based on facts observed at first hand among savages, to discuss the first problem—the dependence of the nuclear complex upon the constitution of the family. The treatment of the second problem is reserved for Part II, while in the last two parts the same twin subjects are discussed in a general manner.

2

THE FAMILY IN FATHER-RIGHT AND MOTHER-RIGHT

The best way to examine this first problem—in what manner the 'family complex' is influenced and modified by the constitution of the family in a given society—is to enter concretely into the matter, to follow up the formation of the complex in the course of typical family life, and to do it comparatively in the case of different civilizations. I do not propose here to survey all forms of human family, but shall compare in detail two types, known to me from personal observation: the patrilineal family of modern civilization, and the matrilineal family of certain island communities in North-Western Melanesia. These two cases, however, represent perhaps the two most radically different types of family known to sociological observation, and will thus serve our purpose well. A few words will be necessary to introduce the Trobriand Islanders of North-Eastern New Guinea (or North-Western Melanesia) who will form the other term of our comparison, besides our own culture.

These natives are matrilineal, that is, they live in a social order in which kinship is reckoned through the mother only, and succession and inheritance descend in the female line. This means that the boy or girl belongs to the mother's family, clan and community: the boy succeeds to the dignities and social position of the mother's brother, and it is not from the father but from the maternal uncle or maternal aunt, respectively, that a child inherits its possessions.

Every man and woman in the Trobriands settles down eventually to matrimony, after a period of sexual play in childhood, followed by general licence in adolescence, and later by a time when the lovers live together in a more permanent intrigue, sharing with two or three other couples a communal 'bachelor's house'. Matrimony, which is usually monogamous, except with chiefs, who have several wives, is a permanent union, involving sexual exclusiveness, a common economic existence, and an independent household. At first glance it might appear to a superficial observer to be the exact pattern of marriage among ourselves. In reality, however, it is entirely different. To begin with, the husband is not regarded as the father of the children in the sense in which we use this word; physiologically he has nothing to do with their birth, according to the ideas of the natives, who are ignorant of physical fatherhood. Children, in native belief, are inserted into the mother's womb as tiny spirits, generally by the agency of the spirit of a deceased kinswoman of the mother.[1] Her husband has then to protect and cherish the children, to 'receive them in his arms' when they are born, but they are not 'his' in the sense that he has had a share in their procreation.

The father is thus a beloved, benevolent friend, but not a recognized kinsman of the children. He is a stranger, having authority through his personal relations to the child, but not

---

[1] See the writer's *The Father in Primitive Psychology* (Psyche Miniatures), 1927, and 'Baloma, Spirits of the Dead', *Journ. R. Anthrop. Inst.*, 1916.

through his sociological position in the lineage. Real kinship, that is identity of substance, 'same body', exists only through the mother. The authority over the children is vested in the mother's brother. Now this person, owing to the strict taboo which prevents all friendly relations between brothers and sisters, can never be intimate with the mother, or therefore with her household. She recognizes his authority, and bends before him as a commoner before a chief, but there can never be tender relations between them. Her children are, however, his only heirs and successors, and he wields over them the direct *potestas*. At his death his worldly goods pass into their keeping, and during his lifetime he has to hand over to them any special accomplishment he may possess—dances, songs, myths, magic and crafts. He also it is who supplies his sister and her household with food, the greater part of his garden produce going to them. To the father, therefore, the children look only for loving care and tender companionship. Their mother's brother represents the principle of discipline, authority, and executive power within the family.[1]

The bearing of the wife towards her husband is not at all servile. She has her own possessions and her own sphere of influence, private and public. It never happens that the children see their mother bullied by the father. On the other hand, the father is only partially the bread-winner, and has to work mainly for his own sisters, while the boys know that when they grow up they in turn will have to work for their sisters' households.

Marriage is patrilocal: that is, the girl goes to join her husband in his house and migrates to his community, if she comes from another, which is in general the case. The children therefore grow up in a community where they are legally strangers, having

[1] For an account of the strange economic conditions of these natives, see the writer's 'Primitive Economics' in *Economic Journal*, 1921, and *Argonauts of the Western Pacific*, chapters ii and vi. The legal side has been fully discussed in *Crime and Custom in Savage Society*, 1926.

no right to the soil, no lawful pride in the village glory; while their home, their traditional centre of local patriotism, their possessions, and their pride of ancestorship are in another place. Strange combinations and confusion arise, associated with this dual influence.

From an early age boys and girls of the same mother are separated in the family, owing to the strict taboo which enjoins that there shall be no intimate relations between them, and that above all any subject connected with sex should never interest them in common. It thus comes about that though the brother is really the person in authority over the sister, the taboo forbids him to use this authority when it is a question of her marriage. The privilege of giving or withholding consent, therefore, is left to the parents, and the father—her mother's husband—is the person who has most authority, in this one matter of his daughter's marriage.

The great difference in the two family types which we are going to compare is beginning to be clear. In our own type of family we have the authoritative, powerful husband and father backed up by society.[1] We have also the economic arrangement whereby he is the bread-winner, and can—nominally at least—withhold supplies or be generous with them at his will. In the Trobriands, on the other hand, we have the independent mother

[1] I should like to mention that although under 'our own' civilization I am here speaking about the European and American communities in general, I have in mind primarily the average type of continental family, as this was the material on which the conclusions of psycho-analysis were founded. Whether among the higher social strata of the Western European or of the North American cities we are now slowly moving towards a condition of mother-right more akin to the legal ideas of Melanesia than to those of Roman Law and of continental custom, I do not dare to prophesy. If the thesis of this book be correct, some modern developments in matters of sex ('petting parties', etc.), as well as the weakening of the patriarchal system, should deeply modify the configurations of the sentiments within the family.

and her husband, who has nothing to do with the procreation of the children, and is not the bread-winner, who cannot leave his possessions to the children, and has socially no established authority over them. The mother's relatives on the other hand are endowed with very powerful influence, especially her brother, who is the authoritative person, the producer of supplies for the family, and whose possessions the sons will inherit at his death. Thus the pattern of social life and the constitution of the family are arranged on entirely different lines from those of our culture.

It might appear that while it would be interesting to survey the family life in a matrilineal society, it is superfluous to dwell on our own family life, so intimately known to everyone of us and so frequently recapitulated in recent psycho-analytic literature. We might simply take it for granted. But first of all, it is essential in a strict comparative treatment to keep the terms of the comparison clearly before our eyes; and then, since the matrilineal data to be given here have been collected by special methods of anthropological field work, it is indispensable to cast the European material into the same shape, as if it had been observed by the same methods and looked at from the anthropological point of view. I have not, as stated already, found in any psycho-analytic account any direct and consistent reference to the social milieu, still less any discussion of how the nuclear complex and its causes vary with the social stratum in our society. Yet it is obvious that the infantile conflicts will not be the same in the lavish nursery of the wealthy bourgeois as in the cabin of the peasant, or in the one-room tenement of the poor working-man. Now just in order to vindicate the truth of the psycho-analytic doctrine, it would be important to consider the lower and the ruder strata of society, where a spade is called a spade, where the child is in permanent contact with the parents, living and eating in the same room and sleeping in the same bed, where no 'parent substitute' complicates the picture, no good manners modify the brutality of the impact, and where the jealousies and petty competitions of daily life clash in hardened though repressed hostility.[1]

It may be added that when we study the nuclear complex and its bed-rock of social and biological actuality in order to apply it to the study of folklore, the need of not neglecting the peasant and the illiterate classes is still more urgent. For the popular traditions originated in a condition more akin to that of the modern Central and Eastern European peasant, or of the poor artisan, than to that of the overfed and nervously overwrought people of modern Vienna, London, or New York.

In order to make the comparison stand out clearly I shall divide the history of childhood into periods, and treat each of them separately, describing and comparing it in both societies. The clear distinction of stages in the history of family life is important in the treatment of the nuclear complex, for psycho-analysis—and here really lies one of its chief merits—has brought to light the stratification of the human mind, and shown its rough correspondence to the stages in the child's development. The distinct periods of sexuality, the crises, the accompanying repressions and amnesias in which some memories are relegated to the unconscious—all these imply a clear division of the child's life into periods.[2] For the present purpose

[1] My personal knowledge of the life, customs and psychology of Eastern European peasants has allowed me to ascertain deep differences between the illiterate and the educated classes of the same society as regards the mental attitude of parents to children and vice versa.

[2] Although in Professor Freud's treatment of infantile sexuality, the division into several distinct stages plays a fundamental rôle, yet in his most detailed work on the subject (*Drei Abhandlungen zur Sexualtheorie,* 5th edition) the scheme of the successive stages is not lucidly nor even explicitly drawn up. This makes the reading of this book somewhat difficult to a non-specialist in psycho-analysis, and it creates certain ambiguities and contradictions, real and apparent, which the present writer has not yet fully solved. Flügel's otherwise excellent exposition of psycho-analysis (*op. cit.*) also suffers from this defect, especially

it will be enough to distinguish four periods in the development of the child, defined by biological and sociological criteria.

1. *Infancy*, in which the baby is dependent for its nourishment on the mother's breast and for safety on the protection of the parent, in which he cannot move independently nor articulate his wishes and ideas. We shall reckon this period as ranging from birth to the time of weaning. Among savage peoples, this period lasts from about two to three years. In civilized communities it is much shorter—generally about one year only. But it is better to take the natural landmarks to divide the stages of childhood. The child is at this time physiologically bound up with the family.

2. *Babyhood*, the time in which the offspring, while attached to the mother and unable to lead an independent existence, yet can move, talk, and freely play round about her. We shall reckon this period to take up three or four years, and thus bring the child to the age of about six. This term of life covers the first gradual severing of the family bonds. The child learns to move away from the family and begins to be self-sufficient.

3. *Childhood*, the attainment of relative independence, the epoch of roving about and playing with other children. This is the time also when in all branches of humanity and in all classes of a society the child begins in some way or other to become initiated into full membership of the community. Among some savages, the preliminary rites of initiation begin. Among others and among our own peasants and working people, especially on the Continent, the child begins to be apprenticed to his future economic life. In Western European and American communities children begin their schooling at this time. This is the period of

---

regrettable in a work which sets out to clear up and systematize the doctrine. The word 'child' throughout the book is used sometimes to mean 'baby', sometimes 'adolescent', and the sense as a rule has to be inferred from the context. In this respect I hope that the present outline will be of some use.

the second severing from family influences, and it lasts till puberty, which forms its natural term.

4. *Adolescence*, between physiological puberty and full social maturity. In many savage communities, this epoch is encompassed by the principal rites of initiation, and in other tribes it is the epoch in which tribal law and order lay their claim on the youth and on the maiden. In modern, civilized communities it is the time of secondary and higher schooling, or else of the final apprenticeship to the life task. This is the period of complete emancipation from the family atmosphere. Among savages and in our own lower strata it normally ends with marriage and the foundation of a new family.

3

THE FIRST STAGE OF
THE FAMILY DRAMA

It is a general characteristic of the mammals that the offspring is not free and independent at birth, but has to rely for its nourishment, safety, warmth, cleanliness and bodily comfort on the care of its mother. To this correspond the various bodily arrangements of mother and child. Physiologically there exists a passionate instinctive interest of the mother in the child, and a craving of the suckling for the maternal organism, for the warmth of her body, the support of her arms and, above all, the milk and contact of her breast. At first the relation is determined by the mother's selective passion—to her only her own off-spring is dear, while the baby would be satisfied with the body of any lactant woman. But soon the child also distinguishes, and his attachment becomes as exclusive and individual as that of the mother. Thus birth establishes a link for life between mother and child.

This link is first founded on the biological fact that young mammals cannot live unaided, and thus the species depends for its survival on one of the strongest instincts, that of maternal love. But society hastens to step in and to add its at first feeble decree to the powerful voice of nature. In all human communities, savage or civilized, custom, law and morals, sometimes even religion, take cognizance of the bond between mother and offspring, usually at as early a stage as the beginning of gestation. The mother, sometimes the father also, has to keep various taboos and observances, or perform rites which have to do with the welfare of the new life within the womb. Birth is always an important social event, round which cluster many traditional usages, often associated with religion. Thus even the most natural and most directly biological tie, that between mother and child, has its social as well as its physiological determination, and cannot be described without reference to the influence exercised by the tradition and usage of the community.

Let us briefly summarize and characterize these social co-determinants of motherhood in our own society. Maternity is a moral, religious and even artistic ideal of civilization, a pregnant woman is protected by law and custom, and should be regarded as a sacred object, while she herself ought to feel proud and happy in her condition. That this is an ideal which can be realized is vouched for by historical and ethnographical data. Even in modern Europe, the orthodox Jewish communities of Poland keep it up in practice, and amongst them a pregnant woman is an object of real veneration, and feels proud of her condition. In the Christian Aryan societies, however, pregnancy among the lower classes is made a burden, and regarded as a nuisance; among the well-to-do people it is a source of embarrassment, discomfort, and temporary ostracism from ordinary social life. Since we thus have to recognize the importance of the mother's pre-natal attitude for her future sentiment towards her offspring, and since this attitude varies greatly with the milieu and depends on social values, it is important that this sociological problem should be studied more closely.

At birth, the biological patterns and the instinctive impulses of the mother are endorsed and strengthened by society, which, in many of its customs, moral rules and ideals, makes the mother the nurse of the child, and this, broadly speaking, in the low as in the high strata of almost all nations of Europe. Yet even here in a relation so fundamental, so biologically secured, there are certain societies where custom and laxity of innate impulses allow of notable aberrations. Thus we have the system of sending the child away for the first year or so of its life to a hired foster mother, a custom once highly prevalent in the middle classes of France; or the almost equally harmful system of protecting the woman's breasts by hiring a foster mother, or by feeding the child on artificial food, a custom once prevalent among the wealthy classes, though today generally stigmatized as unnatural. Here again the sociologist has to add his share in order to give the true picture of motherhood, as it varies according to national, economic and moral differences.

Let us now turn to consider the same relation in a matrilineal society on the shores of the Pacific. The Melanesian woman shows invariably a passionate craving for her child, and the surrounding society seconds her feelings, fosters her inclinations and idealizes them by custom and usage. From the first moments of pregnancy, the expectant mother is made to watch over the welfare of her future offspring by keeping a number of food taboos and other observances. The pregnant woman is regarded by custom as an object of reverence, an ideal which is fully realized by the actual behaviour and feelings of these natives. There exists an elaborate ceremony performed at the first pregnancy, with an intricate and somewhat obscure aim, but emphasizing the importance of the event and conferring on the pregnant woman distinction and honour.

After the birth, mother and child are secluded for about a

month, the mother constantly tending her child and nursing it, while certain female relatives only are admitted into the hut. Adoption under normal circumstances is very rare, and even then the child is usually given over only after it has been weaned, nor is it ever adopted by strangers, but by nearest relatives exclusively. A number of observances, such as ritual washing of mother and child, special taboos to be kept by the mother, and visits of presentation, bind mother and child by links of custom superimposed upon the natural ones.[1]

Thus in both societies, to the biological adjustment of instinct there are added the social forces of custom, morals and manners, all working in the same direction of binding mother and child to each other, of giving them full scope for the passionate intimacy of motherhood. This harmony between social and biological forces ensures full satisfaction and the highest bliss. Society co-operates with nature to repeat the happy conditions in the womb, broken by the trauma of birth. Dr. Rank, in a work which has already proved of some importance for the development of psycho-analysis,[2] has indicated the significance for later life of intra-uterine existence and its memories. Whatever we might think about the 'trauma' of birth, there is no doubt that the first months after birth realize, by the working of both biological and sociological forces, a state of bliss broken by the 'trauma' of weaning. The exceptional aberrations from this state of affairs are to be found only among the higher strata of civilized communities.

[1] An important form of the taboo observed by a mother after birth is the sexual abstinence enjoined upon her. For a beautiful expression of the high moral view of natives concerning this custom see *The Contact of Races and Clash of Culture*, by G. Pitt-Rivers, 1927, chap. viii, sec. 3.

[2] *Das Trauma der Geburt* (1924). Needless to say, the conclusions of Dr. Rank's book are entirely unacceptable to the present writer, who is not able to adopt any of the recent developments of psycho-analysis nor even to understand their meaning.

We find a much greater difference in the fatherhood of the patriarchal and matrilineal family at this period, and it is rather unexpected to find that in a savage society, where the physical bonds of paternity are unknown, and where mother right obtains, the father should yet stand in a much more intimate relation to he children than normally happens among ourselves. For in our own society, the father plays a very small part indeed in the life of a young infant. By custom, usage, and manners, the well-to-do father is kept out of the nursery, while the peasant or working man has to leave the child to his wife for the greater part of the twenty-four hours. He may perhaps resent the attention which the infant claims, and the time which it takes up, but as a rule he neither helps nor interferes with a small child.

Among the Melanesians 'fatherhood', as we know, is a purely social relation. Part of this relation consists in his duty towards his wife's children; he is there 'to receive them into his arms', a phrase we have already quoted; he has to carry them about when on the march the mother is tired, and he has to assist in the nursing at home. He tends them in their natural needs, and cleanses them, and there are many stereotyped expressions in the native language referring to fatherhood and its hardships, and to the duty of filial gratitude towards him. A typical Trobriand father is a hard-working and conscientious nurse and in this he obeys the call of duty, expressed in social tradition. The fact is, however, that the father is always interested in the children, sometimes passionately so, and performs all his duties eagerly and fondly.

Thus, if we compare the patriarchal and the matrilineal relation at this early stage, we see that the main point of difference lies with the father. In our society, the father is kept well out of the picture, and has at best a subordinate part. In the Trobriands, he plays a much more active rôle, which is important above all because it gives him a far greater scope for forming ties of affection with his children. In both societies there is found, with a few exceptions, little room for conflict between the biological trend and the social conditions.

4

FATHERHOOD IN MOTHER-RIGHT

We have now reached the period when the child is already weaned, when it is learning to walk and begins to speak. Yet biologically it is but slowly gaining its independence from the mother's body. It clings to her with undiminished, passionate desire for her presence, for the touch of her body and the tender clasp of her arms.

This is the natural, biological tendency, but in our society, at one stage or another, the child's desires are crossed and thwarted. Let us first realize that the period upon which we now enter is introduced by the process of weaning. By this the blissful harmony of infantile life is broken, or at least modified. Among the higher classes, weaning is so prepared, graduated and adjusted that it usually passes without any shock. But among women of the lower classes, in our society, weaning is often a painful wrench for the mother and certainly for the child. Later on, other obstacles tend to obtrude upon the intimacy of the mother with the child, in whom at that stage a notable change is taking place. He becomes independent in his movements, can feed himself, express some of his feelings and ideas, and begins to understand and to observe. In the higher classes, the nursery arrangements separate the mother from the child in a gradual manner. This dispenses with any shock, but it leaves a gap in the child's life, a yearning and an unsatisfied need. In the lower classes, where the child shares the parents' bed, it becomes at a certain time a source of embarrassment and an encumbrance, and suffers a rude and more brutal repulsion.

How does savage motherhood on the coral islands of New Guinea compare at this stage with ours? First of all, weaning takes place much later in life, at a time when the child is already independent, can run about, eat practically everything and follow other interests. It takes place, that is, at a moment when the child neither wants nor needs the mother's breast any more, so that the first wrench is eliminated.

'Matriarchate', the rule of the mother, does not in any way entail a stern, terrible mother-virago. The Trobriand mother carries her children, fondles them, and plays with them at this stage quite as lovingly as at the earlier period, and custom as well as morals expects it from her. The child is bound to her, also, according to law, custom and usage, by a closer tie than is her husband, whose rights are subservient to those of the offspring. The psychology of the intimate marital relations has therefore a different character, and the repulsion of the child from the mother by the father is certainly not a typical occurrence, if it ever occurs at all. Another difference between the Melanesian and the typical European mother is that the former is much more indulgent. Since there is little training of the child, and hardly any moral education; since what there is begins later and is done by other people, there is scarcely room for severity. This absence of maternal discipline precludes on the one hand such aberrations of severity as are sometimes found among us; on the other hand, however, it lessens the feeling of interest on the part of the child, the desire to please the mother, and to win her approval. This desire, it must be remembered, is one of the strong links of filial attachment among us, and one which holds great possibilities for the establishment of a permanent relation in later life.

Turning now to the paternal relation we see that, in our society, irrespective of nationality or social class, the father still enjoys the patriarchal status.[1] He is the head of the family and the relevant link in the lineage, and he is also the economic provider.

As an absolute ruler of the family, he is liable to become a tyrant, in which case frictions of all sort arise between him and his wife and children. The details of these depend greatly on the social milieu. In the wealthy classes of Western civilization, the child is well separated from his father by all sorts of nursery arrangements. Although constantly with the nurse, the child is usually attended to and controlled by the mother, who, in such cases, almost invariably takes the dominant place in the child's affections. The father, on the other hand, is seldom brought within the child's horizon, and then only as an onlooker and stranger, before whom the children have to behave themselves, show off and perform. He is the source of authority, the origin of punishment, and therefore becomes a bogey. Usually the result is a mixture; he is the perfect being for whose benefit everything has to be done; and, at the same time, he is the 'ogre' whom the child has to fear and for whose comfort, as the child soon realizes, the household is arranged. The loving and

---

[1] Here again I should like to make an exception with regard to the modern American and British family. The father is in process of losing his patriarchal position. As conditions are in flux, however, it is not safe to take them into consideration here. Psycho-analysis cannot hope, I think, to preserve its 'Œdipus complex' for future generations, who will only know a weak and henpecked father. For him the children will feel indulgent pity rather than hatred and fear!

sympathetic father will easily assume the former rôle of a demi-god. The pompous, wooden, or tactless one will soon earn the suspicion and even hate of the nursery. In relation to the father, the mother becomes an intermediary who is sometimes ready to denounce the child to the higher authority, but who at the same time can intercede against punishment.

The picture is different, though the results are not dissimilar, in the one-room and one-bed households of the poor peasantry of Central and Eastern Europe, or of the lower working classes. The father is brought into closer contact with the child, which in rare circumstances allows of a greater affection, but usually gives rise to much more acute and chronic friction. When a father returns home tired from work, or drunk from the inn, he natur-ally vents his ill-temper on the family, and bullies mother and children. There is no village, no poor quarter in a modern town, where cases could not be found of sheer, patriarchal cruelty. From my own memory, I could quote numerous cases where peasant fathers would, on returning home drunk, beat the chil-dren for sheer pleasure, or drag them out of bed and send them into the cold night.

Even at best, when the working father returns home, the chil-dren have to keep quiet, stop rowdy games and repress spon-taneous, childish outbursts of joy and sorrow. The father is a supreme source punishment in poor households also, while the mother acts as intercessor, and often shares in the treatment meted out to the children. In the poorer households, moreover, the economic rôle of provider and the social power of the father are more quickly and definitely recognized, and act in the same direction as his personal influence.

The rôle of the Melanesian father at this stage is very different from that of the European patriarch. I have briefly sketched in Chapter 4 his very different social position as husband and father, and the part he plays in the household. He is not the head of the family, he does not transmit his lineage to his children, nor is he the main provider of food. This entirely changes his legal rights and his personal attitude to his wife. A Trobriand man will seldom quarrel with his wife, hardly ever attempt to brutalize her, and he will never be able to exercise a permanent tyranny. Even sexual co-habitation is not regarded by native law and usage as the wife's duty and the husband's privilege, as is the case in our society. The Trobriand natives take the view, dictated by tradition, that the husband is indebted to his wife for sexual services that he has to deserve them and pay for them. One of the ways, the chief way, in fact, of acquitting himself of this duty is by performing services for the children and showing affection to them. There are many native sayings which embody in a sort of loose folklore these principles. In the child's infancy the husband has been the nurse, tender and loving; later on in early child-hood he plays with it, carries it, and teaches it such amusing sports and occupations as take its fancy.

Thus the legal, moral and customary tradition of the tribe and all the forces of organization combine to give the man, in his conjugal and paternal rôle an entirely different attitude from that of a patriarch. And though it has to be defined in an abstract manner, this is by no means a mere legal principle, detached from life. It expresses itself in every detail of daily existence, permeates all the relations within the family, and dominates the sentiments found there. The children never see their mother subjugated or brutalized or in abject dependence upon her husband, not even when she is a commoner married to a chief. They never feel his heavy hand on themselves; he is not their kinsman, nor their owner, nor their benefactor. He has no rights or pre-rogatives. Yet he feels, as does every normal father all over the world, a strong affection for them; and this, together with his traditional duties, makes him try to win their love, and thus to retain his influence over them.

Comparing European with Melanesian paternity, it is import-ant to keep in mind the biological facts as well as the sociological.

Biologically there is undoubtedly in the average man a tendency towards affectionate and tender feelings for his children. But this tendency seems not to be strong enough to outweigh the many hardships which children entail on a parent. When, therefore, society steps in and in one case declares that the father is the absolute master, and that the children should be there for his benefit, pleasure and glory, this social influence tilts the balance against a happy equilibrium of natural affection and natural impatience of the nuisance. When, on the other hand, a matri-lineal society grants the father no privileges and no right to his children's affections, then he must earn them, and when again, in the same uncivilized society, there are fewer strains on his nerves and his ambitions and his economic responsibilities, he is freer to give himself up to his paternal instincts. Thus in our society the adjustment between biological and social forces, which was satisfactory in earliest childhood, begins to show a lack of harmony later on. In the Melanesian society, the harmonious relations persist.

Father-right, we have seen, is to a great extent a source of family conflict, in that it grants to the father social claims and prerogatives not commensurate with his biological propensities, nor with the personal affection which he can feel for and arouse in his children.

5

INFANTILE SEXUALITY

Traversing the same ground as Freud and the psycho-analysts, I have yet tried to keep clear of the subject of sex, partly in order to emphasize the sociological aspect in my account, partly in order to avoid moot theoretical distinctions as to the nature of mother-and-child attachment or the 'libido'. But at this stage, as the children begin to play independently and develop an interest in the surrounding work and people, sexuality makes its first appearance in forms accessible to outside sociological observation and directly affecting family life.[1] A careful observer of European children, and one who has not forgotten his own childhood, has to recognize that at an early age, say,

[1] The reader who is interested in infantile sexuality and child psychology should also consult A. Moll, *Das Sexualleben des Kindes* (1908); Havelock Ellis's *Studies in the Psychology of Sex* (1919. ed., pp.13 *seqq.*, also vol. i, 1910 ed., pp. 36 *seqq.* and 235 *seqq.* and passim). The books of Ploss-Renz, *Das Kind in Brauch und Sitte der Völker* (Leipzig, 1911–12); Charlotte Bühler, *Das Seelenleben des Jugendlichen* (1925); and the works of William Stern on Child Psychology are also important.

between three and four, there arises in them a special sort of interest and curiosity. Besides the world of lawful, normal and 'nice' things, there opens up a world of shame-faced desires, clandestine interests and subterranean impulses. The two categories of things, 'decent' and 'indecent', 'pure' and 'impure', begin to crystallize, categories destined to remain throughout life. In some people the 'indecent' becomes completely suppressed, and the right values of decency become hypertrophied into the virulent virtue of the puritan, or the still more repulsive hypocrisy of the conventionally moral. Or the 'decent' is altogether smothered through glut in pornographic satisfaction, and the other category develops into a complete pruriency of mind, not less repulsive than hypocritical 'virtue' itself.

In the second stage of childhood which we are now considering, that is according to my scheme from an age of about four to six years, the 'indecent' centres round interests in excretory functions, exhibitionism and games with indecent exposure, often associated with cruelty. It hardly differentiates between the sexes, and is little interested in the act of reproduction. Anyone who has lived for a long time among peasants and knows intimately their childhood will recognize that this state of affairs exists as a thing normal, though not open. Among the working classes things seem to be similar.[1] Among the higher classes 'indecencies' are much more suppressed, but not very different. Observations in these social strata, which would be more difficult than among peasants, should, however, be urgently carried out for pedagogical, moral and eugenic reasons, and suitable methods of research devised. The results would, I think, confirm to an

---

[1] That conscientious sociologist, Zola, has provided us with rich material on the subject, entirely in agreement with my own observations.

extraordinary degree some of the assertions of Freud and his school.[1]

How does the newly awakened infantile sexuality or infantile indecency influence the relation to the family? In the division between things 'decent' and 'indecent', the parents, and especially the mother, are included wholly within the first category, and remain in the child's mind absolutely untouched by the 'indecent'. The feeling that the mother might be aware of any prurient infantile play is extremely distasteful to the child, and there is a strong disinclination to allude in her presence or to speak with her about any sexual matters. The father, who is also kept strictly outside the 'indecent' category, is, moreover, regarded as the moral authority whom these thoughts and pastimes would offend. For the 'indecent' always carries with it a sense of guilt.[2]

Freud and the psycho-analytic school have laid great stress on the sexual rivalry between mother and daughter, father and son

[1] Freud's contentions of the normal occurrence of premature sexuality, of little differentiation between the sexes, of anal-eroticism and absence of genital interest are, according to my observations, correct. In a recent article (Zeitschrift für Psycho-Analyse, 1923), Freud has somewhat modified his previous view, and affirms, without giving arguments, that children at this stage have, after all, already a 'genital' interest. With this I cannot agree.

[2] The attitude of the modern man and woman is rapidly changing. At present we studiously 'enlighten' our children, and keep 'sex' neatly prepared for them. In the first place, however, we must remember that we are dealing here with a minority even among the British and American 'intelligentsia'. In the second place, I am not at all certain whether the bashfulness and awkward attitude of children towards their parents in matters of sex will be to any great extent overcome by this method of treatment. There seems to exist a general tendency even among adults to eliminate the dramatic, upsetting, and mysterious emotional elements out of any stable relationship based on every-day intercourse. Even among the essentially 'unrepressed' Trobrianders the parent is never the confidant in matters of sex. It is remarkable how much easier it is to make any delicate or shameful confession to those friends and acquaintances who are not too intimately connected with our daily life.

respectively. My own opinion is that the rivalry between mother and daughter does not begin at this early stage. At any rate, I have never observed any traces of it. The relations between father and son are more complex. Although, as I have said, the little boy has no thoughts, desires or impulses towards his mother which he himself would feel belong to the category of the 'indecent', there can be no doubt that a young organism reacts sexually to close bodily contact with the mother.[1] A well-known piece of advice given by old gossips to young mothers in peasant communities is to the effect that boys above the age of three should sleep separately from the mother. The occurrence of infantile erections is well known in these communities, as is also the fact that the boy clings to the mother in a different way from the girl. That the father and the young male child have a component of sexual rivalry under such conditions seems probable, even to an outside sociological observer. The psycho-analysts maintain it categorically. Among the wealthier classes crude conflicts arise more seldom, if ever, but they arise in imagination and in a more refined though perhaps not less insidious form.

It must be noted that at this stage when the child begins to show a different character and temperament according to sex, the parents' feelings are differentiated between sons and daughters. The father sees in the son his successor, the one who is to replace him in the family lineage and in the household. He becomes therefore all the more critical, and this influences his feelings in two directions: if the boy shows signs of mental or physical deficiency, if he is not up to the type of the ideal in which the father believes, he will be a source of bitter

[1] Since this was first written in 1921, I have changed my views on this subject. The statement that 'a young organism reacts sexually to close bodily contact with the mother' appears to me now absurd. I am glad I may use this strong word, having written the absurd statement myself. I have set forth what appears to me the adequate analysis of this phase in infantile psychology later on, Part IV, Chapter 9.

disappointment and hostility. On the other hand, even at this stage, a certain amount of rivalry, the resentment of future supersession, and the melancholy of the waning generation lead again to hostility. Repressed in both cases, this hostility hardens the father against the son and provokes by reaction a response in hostile feelings. The mother, on the other hand, has no grounds for negative sentiments, and has an additional admiration for the son as a man to be. The father's feeling towards the daughter—a repetition of himself in a feminine form—hardly fails to evoke a tender emotion, and perhaps also to flatter his vanity. Thus social factors mix with biological and make the father cling more tenderly to the daughter than to the son, while with the mother it is the reverse. But it must be noted that an attraction to the offspring of the other sex, because it is of the other sex, is not necessarily sexual attraction.

In Melanesia, we find an altogether different type of sexual development in the child. That the biological impulses do not essentially differ, seems beyond doubt. But I have failed to find any traces of what could be called infantile indecencies, or of a subterranean world in which children indulge in clandestine pastimes centring round excretory functions or exhibitionism. The subject naturally presents certain difficulties of observation, for it is hard to enter into any personal communication with a savage child, and if there were a world of indecent things as amongst ourselves, it would be as futile to inquire about it from an average grown-up native as from a conventional mother, father, or nurse in our society. But there is one circumstance which makes matters so entirely different among these natives that there is no danger of making a mistake: this is that with them there is no repression, no censure, no moral reprobation of infantile sexuality of the genital type when it comes to light at a somewhat later stage than the one we are now considering—at about the age of five or six. So if there were any earlier indecency, this could be as easily observed as the later genital stage of sexual plays.

How can we then explain why among savages there is no period of what Freud calls 'pre-genital', 'anal-erotic' interest? We shall be able to understand this better when we discuss the sexuality of the next stage in the child's development, a sexuality in which native Melanesian children differ essentially from our own.

6

APPRENTICESHIP TO LIFE

We enter now on the third stage of childhood, commencing between the ages of five and seven. At this period a child begins to feel independent, to create its own games, to seek for associates of the same age, with whom it has a tendency to roam about unencumbered by grown-up people. This is the time when play begins to pass into more definite occupations and serious life interests.

Let us follow our parallel at this stage. In Europe, entrance into school or, among the uneducated classes, some sort of preliminary apprenticeship to an economic occupation removes the child from the influence of the family. The boy or girl lose to some extent their exclusive attachment to the mother. With the boy, there frequently takes place at this period a transference of sentiment to a substitute mother, who for the time being is regarded with some of the passionate tenderness felt for the mother, but with no other feelings. Such transference must not be confused with the much later tendency of adolescent boys to fall in love with women older than themselves. At the same time, there arises a desire for independence from the all-possessive intimacy of maternal interest, which makes the child withhold its absolute confidence from the parent. Among the peasants and lower classes, the process of emancipation from the mother takes place earlier than in the higher, but it is similar in all essentials. When the mother is deeply attached to the child, especially to the boy, she is apt to feel a certain amount of jealousy and resentment at this emancipation and to put obstacles in its way. This usually makes the wrench only more painful and violent.

The children on the coral beaches of the Western Pacific show a similar tendency. This appears there even more clearly, for the absence of compulsory education and of any strict discipline at this age allows a much freer play to the natural inclinations of infantile nature. On the part of the mother there is in Melanesia, however, no jealous resentment or anxiety at the child's new-found independence, and here we see the influence of the lack of any deep, educational interest between mother and child. At this stage, the children in the Trobriand archipelago begin to form a small juvenile community within the community. They roam about in bands, play on distant beaches or in secluded parts of the jungle, join with other small communities of children from neighbouring villages, and in all this, though they obey the commands of their child-leaders, they are almost completely independent of the elders' authority. The parents never try to keep them back, to interfere with them in any way or to bind them to any routine. At first, of course, the family still retains a considerable hold over the child, but the process of emancipation progresses gradually and constantly in an untrammelled, natural manner.

In this there is a great difference between European conditions, where the child often passes from the intimacy of the family to the cold discipline of the school or other preliminary training, and the Melanesian state of affairs where the process of emancipation is gradual, free and pleasant.

And now what about the father at this stage? In our society—here again excluding certain modern phases of family life in Britain and America—he still represents the principle of authority within the family. Outside, at school, in the workshop, at the preliminary manual labour which the child of peasants is often set to do, it is either the father in person or his substitute who wields the power. In the higher classes at this stage, the very important process of conscious formation of paternal authority and of the father ideal takes place. The child begins now to comprehend what it had guessed and felt before—the father's established authority as the head of the family, and his economic influence. The ideal of his infallibility, wisdom, justice, and might is usually in varying degrees and in different ways inculcated in the child by the mother or the nurse in religious and moral teaching. Now the rôle of an ideal is never an easy one, and to maintain it in the intimacy of daily life is a very difficult performance indeed. especially for one whose bad tempers and follies are not repressed by any discipline. Thus no sooner is the father ideal formed than it begins to decompose. The child feels at first only a vague malaise at his father's bad temper or weakness, a fear of his wrath, a dim feeling of injustice, perhaps some shame when the father has a really bad outburst. Soon the typical father-sentiment is formed, full of contradictory emotions, a mixture of reverence, contempt, affection and dislike, tenderness and fear. It is at this period of childhood that the social influence due to patriarchal institutions makes itself felt in the child's attitude towards the male parent. Between the boy and his father the rivalries of successor and superseded, and the mutual jealousies described in the previous section, crystallize more distinctly and make the negative elements of the father-to-son relation more predominant than in the case of father-to-daughter.

Among the lower classes, the process of the idealization of the father is cruder but not less important. As I have already said, the father in a typical peasant household is openly a tyrant. The mother acquiesces in his supremacy and imparts the attitude to her children, who reverence and at the same time fear the strong and brutal force embodied in their father. Here also a sentiment composed of ambivalent emotion is formed, with a distinct preference of the father for his female children.

What is the father's rôle in Melanesia? Little need be said about it at this stage. He continues to befriend the children, to help them, to teach them what they like and as much as they like. Children, it is true, are less interested in him at this stage and prefer, on the whole, their small comrades. But the father is always there as a helpful adviser, half playmate, half protector.

Yet at this period the principle of tribal law and authority, the submission to constraint and to the prohibition of certain desirable things enters the life of a young girl or boy. But this law and constraint are represented by quite another person than the father, by the mother's brother, the male head of the family in a matriarchal society. He it is who actually wields the *potestas* and who indeed makes ample use of it.

His authority, though closely parallel to that of the father among ourselves, is not exactly identical with it. First of all his influence is introduced into the child's life much later than that of the European father. Then again, he never enters the intimacy of family life, but lives in another hut and often in a different village, for, since marriage is patrilocal in the Trobriands, his sister and her children have their abode in the village of the husband and father. Thus his power is exercised from a distance and it cannot become oppressive in those small matters which are most irksome. He brings into the life of the child, whether boy or girl, two elements: first of all, that of duty, prohibition and constraint: secondly, especially into the life of the boy, the elements of ambition, pride and social values, half of that, in fact, which makes life worth living for the Trobriander. The constraint comes in, in so far as he begins to direct the boy's occupations, to require certain of his services and to teach him some of the tribal laws and prohibitions. Many of these have already been inculcated into the boy by the parents, but the *kada* (mother's brother) is always held up to him as the real authority behind the rules.

A boy of six will be solicited by his mother's brother to come on an expedition, to begin some work in the gardens, to assist in the carrying of crops. In carrying out these activities, in his maternal uncle's village and together with other members of his clan, the boy learns that he is contributing to the *butura* of his clan; he begins to feel that this is his own village and own people; to learn the traditions, myths and legends of his clan. The child at this stage also frequently co-operates with his father, and it is interesting to note the difference in the attitude he has toward the two elders. The father still remains his intimate; he likes to work with him, assist him and learn from him; but he realizes more and more that such co-operation is based on goodwill and not on law, and that the pleasure derived from it must be its own reward, but that the glory of it goes to a clan of strangers. The child also sees his mother receiving orders from her brother, accepting favours from him, treating him with the greatest reverence, crouching before him like a commoner to a chief. He gradually begins to understand that he is his maternal uncle's successor, and that he will also be a master over his sisters, from whom at this time he is already separated by a social taboo forbidding any intimacy.

The maternal uncle is, like the father among us, idealized to the boy, held up to him as the person who should be pleased, and who must be made the model to be imitated in the future. Thus we see that most of the elements, though not all, which make the father's rôle so difficult in our society, are vested among the Melanesians in the mother's brother. He has the power, he is idealized, to him the children and the mother are subjected, while the father is entirely relieved of all these odious

prerogatives and characteristics. But the mother's brother introduces the child to certain new elements which make life bigger,
more interesting, and of greater appeal—social ambition, traditional glory, pride in his lineage and kinship, promises of future
wealth, power, and social status.

It must be realized that at the time when our European child
starts to find its way in our complex social relations, the Melanesian girl or boy also begins to grasp the principle of kinship
which is the main foundation of the social order. These principles cut across the intimacy of family life and rearrange for the
child the social world which up to now consisted for him of the
extended circles of family, further family, neighbours and village
community. The child now learns that he has to distinguish
above and across these groups' two main categories. The one
consists of his real kinsmen, his *veyola*. To these belong in the first
place his mother, his brothers and sisters, his maternal uncle and
all their kinsmen. These are people who are of the same substance or the 'same body' as himself. The men he has to obey, to
co-operate with and to assist in work, war and personal quarrels.
The women of his clan and of his kinship are strictly tabooed
sexually for him. The other social category consists of the
strangers or 'outsiders', *tomakava*. By this name are called all those
people who are not related by matrilineal ties, or who do not
belong to the same clan. But this group comprises also the father
and his relations, male and female, and the women whom he
may marry or with whom he may have love affairs. Now these
people, and especially the father, stand to him in a very close
personal relation which, however, is entirely ignored by law and
morality. Thus we have on the one side the consciousness of
identity and kinship associated with social ambitions and
pride, but also with constraint and sexual prohibition; and on
the other, in the relation to the father and his relatives, free
friendship and natural sentiment as well as sexual liberty, but no
personal identity or traditional bonds.

7

THE SEXUALITY OF LATER CHILDHOOD

We pass now to the problem of sexual life in the third period—the later childhood, as we might call it, covering the stage of free play and movement, which lasts from about five or six till puberty. I kept the discussion of sex separate from that of the social influences when dealing with the previous period of child life, and I shall do the same here, so as to bring out clearly the respective contributions of organism and society.

In modern Europe, according to Freud, there sets in at this age a very curious phenomenon: the regression of sexuality, a period of latency, a lull in the development of sexual functions and impulses. What makes this latency period especially important in the Freudian scheme of neuroses is the amnesia which is associated with it, the curtain of complete oblivion which falls at this period and which obliterates the reminiscences of infantile sexuality. Remarkably enough, this important and interesting contention of Freud's is not endorsed by other students. For instance, Moll, in his memoir on infantile sexuality (a very thorough and competent contribution),[1] makes no mention of any lull in sexual development. On the contrary, his account implies a steady and gradual increase of sexuality in the child, the curve rising in a continuous manner without any kink. It is remarkable to find that Freud himself at times appears to vacillate. Thus of all the periods of childhood this one has no clear and explicit chapter devoted to it and in one or two places Freud even withdraws his contention about its existence.[2] Yet, if I may affirm on the basis of material derived from personal knowledge of well-educated schoolboys, the latency period invariably sets in at about the sixth year and lasts from two to four years. During this time interest in indecencies flags, the lurid yet alluring colours which they had fade away, and they are repressed and forgotten while new things arise to take up the interest and energy.

How are we to explain the divergency in Freud's own views as well as the ignoring of the facts by other students of sex?

It is clear that we do not deal here with a phenomenon deeply rooted in man's organic nature, but with one largely if not wholly determined by social factors. If we turn to a comparative survey of the various layers of society, we perceive without difficulty that among the lower classes, especially among peasants, the latency period is much less pronounced. In order to see matters clearly, let us cast back to the previous period of infantile pregenital sexuality and see how the two link up. We saw in Chapter V that in the lower as well as in the higher strata there exists at an early age this strong interest in the 'indecent'. Among peasant children, however, it appears somewhat later and has a slightly different character. Let us compare once more the

---

[1] A. Moll, *Das Sexualleben des Kindes*, 1908.

[2] The latency period is frequently mentioned, for instance, in *Drei Abhandlungen*, 5th edition, pp. 40, 44, 64; *Vorlesungen*, 1922, p. 374. But there is no special treatment of it in any of these books. Again, we read, 'Die Latenzzeit kann auch entfallen. Sie braucht keine Unterbrechung der Sexualbetätigung, der Sexualinteressen mit sich zu bringen', *Vorlesungen, loc. cit.*

sources of 'anal-eroticism', as it is called by Freud, among the children of the lower and higher classes.[1] In the nursery of the well-to-do baby, the natural functions, the interest in excretion, are at first encouraged, and then suddenly stopped. The nurse or mother, who up to a certain point tries to stimulate the perform- ance, praises the prompt execution and shows the results, dis- covers at a certain moment that the child takes too much interest in it and begins to play in a manner that to the grown-up appears unclean, though to the child it is perfectly natural. Then the nursery authority steps in, slaps the child, makes it an offence, and the interest is violently repressed. The child grows up, the reticences, frowns, and artificialities begin to surround the nat- ural functions with clandestine interest and mysterious attraction.

Those who remember from their own childhood how strongly such a repressive atmosphere of hints and *sous-entendus* is felt and how well its meaning is understood by the child, recog- nize that the category of 'indecent' is created by elders. From observations of children, moreover, as well as from memory, it is easy to ascertain how quickly and how soon the children catch up artificial attitudes of elders, becoming little prigs, moralists, and snobs. Among peasants, conditions are quite different. The children are instructed in sexual matters at an early age: they cannot help seeing sexual performances of their parents and other relatives; they listen to quarrels in which whole lists of sexual obscenities and technicalities are recited. They have to deal with domestic animals, whose propagation in all its details is a matter of great concern to the whole household and is freely and minutely discussed. Since they are deeply steeped in things natural, they feel less inclined to amuse themselves by doing in a

---

[1] I would not now use the ugly neologism 'anal-eroticism', but as long as a term is defined there is no harm in borrowing it from a doctrine which is being discussed.

clandestine manner that which in many ways they can do and enjoy openly. The children of the working classes stand perhaps midway between the two extremes. Hardly in contact with animals, they receive, on the other hand, an even greater amount of bedroom demonstration and public-house instruction.

What is the result of these essential differences between well-to-do and proletarian children? First of all, the 'indecency' which among bourgeois children is fostered by the repression of the natural curiosities is much less pronounced in the lower classes, and comes up only later where indecency is already associated with ideas of genital sexuality. In the higher classes, when the curiosity about indecencies has played itself out, and with the leaving of the nursery new interests in life crop up, the period of latency now sets in and these new interests absorb the child, while the absence of knowledge which is usual among children of the educated prevents the genital interest setting in so early.

In the lower classes this knowledge and early curiosity in genital matters are present at the same time and establish a continuity, a steady development from the early period to that of full sexual puberty.

The nature of social influences collaborates with these facts to produce a much greater breach of continuity in the life of a well-to-do child. While his whole life up to the age of six was devoted to amusement, he has now suddenly to learn and to do school-work. The peasant child had already previously been helping with the cooking or looking after the younger children, or running after the geese and sheep. At this time, there is no breach of continuity in his life.

Thus, while the early childish interest in the indecent awakens earlier and in another form in the peasant and proletarian child, it is less clandestine, less associated with guilt, hence less immoral, less 'anal-erotic' and more attached to sex. It passes more easily and with more continuity into early sexual play, and the period of latency is almost completely absent or, at any rate, much less pronounced. This explains why psycho-analysis, which deals with neurotic well-to-do people, has led to the discovery of this period, while the general medical observations of Dr. Moll did not detect it.

But if there could be any doubt about the facts of this difference between the classes and about its cause, such doubt should disappear when we turn to Melanesia. Here certainly the facts are different from those found among our educated classes. As we saw in Chapter V, the early sexual indecencies, clandestine games and interests are absent. In fact, it might be said that for these children the categories of decent–indecent, pure–impure, do not exist. The same reasons which make this distinction weaker and less important among our peasants than among our bourgeois act even more strongly and directly among the Melanesians. In Melanesia there is no taboo on sex in general; there is no putting of any veils on natural functions, certainly not in the case of a child. When we consider that these children run about naked, that their excretory functions are treated openly and naturally, that there is no general taboo on bodily parts or on nakedness in general; when we further consider that small children at the age of three and four are beginning to be aware of the existence of such a thing as genital sexuality, and of the fact that this will be their pleasure quite soon just as other infantile plays will be—we can see that social factors rather than biological explain the difference between the two societies.

The stage which I am now describing in Melanesia—that which corresponds to our latency period—is the stage of infantile independence, where small boys and girls play together in a sort of juvenile republic. Now, one of the main interests of these children consists of sexual pastimes. At an early age children are initiated by each other, or sometimes by a slightly older companion, into the practices of sex. Naturally at this stage they are unable to carry out the act properly, but they content themselves with all sorts of games in which they are left quite at liberty by their elders, and thus they can satisfy their curiosity and their sensuality directly and without disguise.

There can be no doubt that the dominating interest of such games is what Freud would call 'genital', that they are largely determined by the desire to imitate the acts and interests of elder children and elders, and that this period is one which is almost completely absent from the life of better-class children in Europe, and which exists only to a small degree among peasants and proletarians. When speaking of these amusements of the children, the natives will frequently allude to them as 'copulation amusement' (mwaygini kwayta). Or else it is said that they are playing at marriage.

It must not be imagined that all games are sexual. Many do not lend themselves at all to it. But there are some particular pastimes of small children in which sex plays the predominant part. Melanesian children are fond of 'playing husband and wife.' A boy and girl build a little shelter and call it their home; there they pretend to assume the functions of husband and wife, and amongst those of course the most important one of sexual intercourse. At other times, a group of children will go for a picnic where the entertainment consists of eating, fighting, and making love. Or they will carry out a mimic ceremonial trade exchange, ending up with sexual activities. Crude sensual pleasure alone does not seem to satisfy them; in such more elaborate games it must be blended with some imaginative and romantic interest.

A very important point about this infantile sexuality is the attitude of the elder generation towards it. As I have said, the parents do not look upon it as in the least reprehensible. Generally they take it entirely for granted. The most they will do is to speak jestingly about it to one another, discussing the love tragedies and comedies of the child world. Never would they dream of interfering or frowning disapproval, provided the children show a due amount of discretion, that is, do not perform their amorous games in the house, but go away somewhere apart in the bush.

But above all the children are left entirely to themselves in their love affairs. Not only is there no parental interference, but rarely, if ever, does it come about that a man or a woman take a perverse sexual interest in children, and certainly they would never be seen to mix themselves up in the games in this rôle. Violation of children is unknown, and a person who played sexually with a child would be thought ridiculous and disgusting.

An extremely important feature in the sexual relations of children is the brother and sister taboo, already mentioned. From an early age, when the girl first puts on her grass petticoat, brothers and sisters of the same mother must be separated from each other, in obedience to the strict taboo which enjoins that there shall be no intimate relations between them. Even earlier, when they first can move about and walk, they play in different groups. Later on they never consort together socially on a free footing, and above all there must never be the slightest suspicion of an interest of one of them in the love affairs of the other. Although there is comparative freedom in playing and language between children, not even quite a small boy would associate sex with his sisters, still less make any sexual allusion or joke in their presence. This continues right through life, and it is the highest degree of bad form to speak to a brother about his sister's love affairs, or vice versa. The imposition of this taboo leads to an early breaking up of family life, since the boys and girls, in order to avoid each other, must leave the parental home and go elsewhere. With all this, we can perceive the enormous difference which obtains in the juvenile sexuality at this stage of later childhood between ourselves and the Melanesians. While amongst ourselves, in the educated classes, there is at this time a break of sexuality and a period of latency with amnesia, in Melanesia the extremely early beginning of genital interest leads to a type of sexuality entirely unknown among us. From this time, the sexuality of the Melanesians will continuously though gradually develop, till it reaches puberty. On the condition that one taboo is respected in the strictest and most complete manner, society gives complete free play to juvenile sexuality.

8

PUBERTY

At an age varying with climate and race and stretching from about the ninth to the fifteenth year, the child enters upon the age of puberty. For puberty is not a moment or a turning point but a more or less prolonged period of development during which the sexual apparatus, the whole system of internal secretions and the organism in general are entirely recast. We cannot consider puberty as a *conditio sine qua non* of sexual interest or even of sexual activities, since non-nubile girls can copulate and immature boys are known to have erections and to practise *immissio penis*. But undoubtedly the age of puberty must be regarded as the most important landmark in the sexual history of the individual.

Sex is, moreover, so intimately bound up at this stage with the other aspects of life that in this chapter we shall treat sexual and social matters together and not divide them as we did in the case of the two previous stages. In comparing here the Trobrianders of Melanesia with our own society, it is important to note that these savages have no initiation rites at puberty. While this will remove one item of extreme importance from our discussion, it will allow us on the other hand to draw the comparison between patriliny and matriliny more clearly and closely, since in most other savage societies initiation ceremonies completely mask or modify this period.

In our own society, we have to speak separately of the boy and of the girl, for at this point the two part company completely in sexual matters. In a man's life, puberty means the acquisition of full mental powers as well as bodily maturity and the final formation of the sexual characters. With his new manliness his whole relation to life in general changes as deeply as his relation to sexual matters and to his position in the family. Beginning with this last, we can observe an extremely interesting phenomenon which greatly affects his attitude to his mother, sister, or other female relatives. The typical adolescent boy of our civilized communities begins to show at the time of puberty an extreme embarrassment towards his mother, affects scorn and a certain brutality towards his sisters and is ashamed before his comrades of all his female relatives. Who of us does not remember the pangs of ineffable shame when, jauntily going along with our school fellows, we met suddenly our mother, our aunt, our sister, or even our girl cousin and were obliged to greet her. There was a feeling of intense guilt, of being caught in *flagrante delicto*. Some boys tried to ignore the embarrassing encounter, others more brave blushed crimson and saluted, but everyone felt that it was a shadow on his social position, an outrage on his manliness and independence. Without entering into the psychology of this phenomenon, we can see that the shame and confusion felt here is of the same type as that associated with any breach of good manners.

This newly acquired manliness affects deeply the boy's attitude towards the world, his whole *Weltanschauung*. He begins to have his independent opinions, his personality and his own honour, to maintain his position towards authority and intellectual leadership. This is a new stage in the relation between father and son, another reckoning up and a new testing of the father ideal. At this point it succumbs if the father is found out to be a fool or a 'bounder', to be a hypocrite or an 'old fogey'. He is usually disposed of for life, and in any case loses the chance of effectively influencing the boy even if in later life the two should come together again. If on the other hand the father can stand the extremely severe scrutiny of this epoch there is a great chance of his surviving as an ideal for life. The reverse is also true, of course, for the father as well keenly examines his son at this epoch, and is equally critical as to whether the boy comes up to his own ideal of what his future successor should be.

The new attitude towards sex, the recrystallization at puberty, exerts a great influence on the boy's attitude not only to his father, but to his mother as well. The educated boy only now fully realizes the biological nature of the bond between his parents and himself. If he deeply loves and worships his mother, as is usually the case, and if he can continue to idealize his father, then the idea of his bodily origin from his parent's sexual intercourse, though at first making a rift in his mental world, can be dealt with. If on the other hand he scorns and hates his father, be it unavowedly as so often happens, the idea brings about a permanent defilement of the mother and a besmirching of things most dear to him.

The new manhood influences above all the boy's sexual outlook. Mentally he is ready for knowledge, physiologically ready for applying it in life. Usually he receives his first lessons in sex at this time, and in some form or other starts sexual activities, not so often, probably, in the normal, regular manner, but frequently through masturbation or nocturnal pollutions. This epoch is in many respects the dividing of the roads for the boy. Either the newly awakened sex impulse, appealing to a strong temperament and to easy morals, absorbs him completely, carries him off his feet once for ever in a wave of over-mastering sensuality;

or else other interests and morality are strong enough to stave it off partially or even completely. As long as he preserves an ideal of chastity and is able to fight for it, the leverage is there for the lifting of the sexual impulses to a higher level. In this, of course, the temptations are largely determined by the social setting and the mode of life of the boy. The racial characteristics of a community, its code of morals and its cultural values establish great differences within European civilization. In certain classes of some countries, it is usual for the boy to succumb to the disintegrating forces of easy sexuality. In others, he can take his chance. In others, again, society relieves him of a great deal of responsibility by laying down rules of stern morality.

In his relations to persons of the other sex, there appears at first something parallel to his attitude towards mother and sister; a certain embarrassment, and polarity of attraction and repulsion. The woman who, he feels, can exercise a deep influence on him alarms him and fills him with suspicion. He senses in her a danger to his awakening independence and manliness.

At this stage also the new fusion of tenderness with sexuality which comes about towards the end of puberty mixes up infantile memories of maternal tenderness with the new elements of sexuality. Imagination and especially dream fantasies bring about horrible confusion and play strange tricks on the boy's mind.[1]

All this refers more especially to the boy belonging to the higher, well-to-do classes. If we compare the peasant or proletarian youth with him, we see that the essential elements are the same, though there is perhaps less individual variation and the general picture is more sober.

Thus there is also a period of affective crudeness towards mother and sister which is especially noticeable in a young peasant. The quarrels with the father crop up as a rule with an

---

[1] This conception is more fully elaborated below, Part IV, Chapter 9.

increased violence now that the boy realizes his own forces and his own position as a successor, now that he feels a new greed for possession and a new ambition for influence. Often a regular fight for supremacy begins at this time. In sexual matters there is not as violent a crisis and this reacts less directly on the parental relation. But the main outlines are the same.

The girl of the educated classes goes through a crisis at her first menstruation which, while it encroaches on liberty and complicates life, adds to its mysterious attraction and is usually eagerly awaited. But puberty is less of a turning point socially to a girl; she continues to live at home or to carry on her education at a boarding school, but all her occupations and her training are in harmony with ordinary family life—not taking the modern, professional girl into account. Her aim in life is to await marriage. One important element in her relation to the family is the rivalry between mother and daughter which often sets in at this time. How often it makes its appearance in a decided, undisguised form[1] is hard to say, but there can be no doubt that it introduces a distorted element into the typical relations of the ordinary family. At this time, also, and not earlier, there enters a special tenderness into the relations between father and daughter, which not infrequently becomes correlated with the maternal rivalry. This is the configuration of the Electra complex; it is therefore of an entirely different nature from the Œdipus complex. Putting on one side the greater hysterical tendency of women, for here we are concerned with the ground work of normality, the Electra complex is less frequent and has less social importance as well as a smaller influence on Western culture. On the other hand, its influence makes itself more frequently felt and the father–daughter incest seems to be incomparably more frequent in real occurrence than that between mother and son,

---

[1] Such as we find it so powerfully described for instance in the very instructive novel of Maupassant, *Fort comme la Mort*.

for various reasons of a biological and sociological nature. Since, however, our interest in this discussion is mainly in the cultural and social influence of the complexes, we cannot follow the parallel between the Œdipus and Electra complexes in detail. Nor can we enter into a comparison between the higher classes, where the repressions are stronger, where there is more hysteria but fewer cases of actual incest occur, and the lower classes, where, since the girl's sex interest is frequently engaged earlier and more normally, she is less liable to hysterical distortions, but suffers more frequently from persecution by the father.[1]

Let us now turn to the Trobriand Islands. Puberty begins there earlier than with us, but at the same time, when it sets in boys and girls have already begun their sexual activities. In the social life of the individual, puberty does not constitute a sharp turning point as in those savage communities where initiation ceremonies exist. Gradually, as he passes to manhood, the boy begins to take a more active part in economic pursuits and tribal occupations, he is considered a young man (ulatile), and by the end of puberty he is a full member of the tribe, ready to marry and carry on all his duties as well as to enjoy his privileges. The girl, who at the beginning of puberty acquires more freedom and independence from her family, has also to do more work, amuse herself more intensely, and carry on such duties, ceremonial, economic, and legal, as are entailed by full womanhood.

But the most important change, and the one which interests us most, is the partial break-up of the family at the time when the adolescent boys and girls cease to be permanent inmates of the parental home. For brothers and sisters, whose avoidance has

---

[1] Among peasants, the attempts of father on daughter are very frequent. This seems especially to be the case among the Latin races. I have been told that in Rumania the occurrence of this type of incest is very common among peasants, and so it seems to be in Italy. In the Canary Islands, I know myself of a few cases of father and daughter committing incest, not in a clandestine manner, but living openly in a shameless ménage and rearing their children.

begun long before in childhood, must now keep an extremely strict taboo, so that any possibility of contact while engaged in sexual pursuits must be eliminated. This danger is obviated by a special institution, the bukumatula. This name is given to special houses inhabited by groups of adolescent boys and girls. A boy as he reaches puberty will join such a house, which is owned by some mature youth or young widower and tenanted by a number of youths, from three to six, who are there joined by their sweethearts.[1] Thus the parental home is drained completely of its adolescent males, though until the boy's marriage he will always come back for food, and will also continue to work for his household to some extent. A girl, on the rare nights of chastity when she is not engaged in one bukumatula or another, may return to sleep at home.

What is the attitude towards mother, father, sister or brother into which the sentiments of the Melanesian boy and girl crystal-lize at this important epoch? As with a modern European boy and girl, we see that at this period there is only a final cast, a consolidation of what has been in gradual formation during the previous stages of life. The mother, from whom the child has been weaned—in the widest sense of the word—remains still the pivotal point of all kinship and relationship for the rest of life. The boy's status in society, his duties and privileges, are determined with regard to her and her relatives. If no one else is there to provide for her, he will have to do it, while her house will always be his second home. Affection and attachment, pre-scribed by social obligations, remain also deeply founded in real sentiment, and when an adult man dies or suffers mishap, his mother will be the one to sorrow and her wailing will last longest and be most sincere. Yet there is little of the personal

---

[1] For a detailed description and analysis of this remarkable institution, as close a mimicry of group-marriage as we have on record, compare the author's forth-coming *Sexual Life of Savages*.

friendship, the mutual confidences and intimacy which is so characteristic of the mother-to-son relationship in our society. The detachment from the mother, carried out as we have seen at every stage more easily and more thoroughly than with us, with fewer premature wrenches and violent suppressions, is achieved in a more complete and harmonious manner.

The father at this time suffers a temporary eclipse. The boy, who as a child was fairly independent and became the member of the small, juvenile republic, gains now on the one hand the additional freedom of the *bukumatula*, while on the other he becomes much more restricted by his various duties towards his *kada* (maternal uncle). He has less time and less interest left for the father. Later on, when friction with the maternal uncle makes its appearance, he turns, as a rule, to his father once more, and their life friendship then becomes settled. At this stage, however, when the adolescent has to learn his duties, to be instructed in traditions and to study his magic, his arts and crafts, his interest in his mother's brother, who is his teacher and tutor, is greatest and their relations are at their best.[1]

There is one more important difference between the Melanesian boy's feeling for his parents and that of the educated boy in our own society. With us, when at puberty and with social initiation the new fiery vision opens before the youth, its glare throws a strange shadow on his previous warm feelings for mother and father. His own sexuality estranges him from his progenitors, embarrasses their relations and creates deep complications. Not so in the matrilineal society. The absence of the early indecency period and of the first struggles against parental

---

[1] The relation between these three, the young man, his father and his mother's brother, are in reality somewhat more complicated than I have been able to show here, and present an interesting picture of the play and clash of the incompatible principles of kinship and authority. This subject will be discussed in a forthcoming book on kinship. Compare also *Crime and Custom*, 1926.

authority; the gradual and open taking-up of sex ever since it first began to stir in the young blood; above all the attitude of benevolent onlookers which the parents take towards the sexuality of their young; the fact of the mother's withdrawing completely but gradually from the boy's passionate feelings; the father smiling his approval—all this brings about the fact that the intensification of sexuality at puberty exercises no direct influence upon the relation to the parents.

One relation, that between brother and sister, is, however, deeply affected by every increase of sexuality—especially at puberty. This taboo, which extends to all free association and excludes the motive of sex completely from the relations of the two, affects the sexual outlook of both in general. For in the first place it must be kept in mind that this taboo is the great sexual barrier in a man's life, beyond which it is illicit to trespass, and that it constitutes also the most important general moral rule. The prohibition, moreover, which starts in childhood with the separation of brothers and sisters and of which this separation always remains the main point, extends also to all other females of the same clan. Thus the sexual world is for the boy divided into two moieties: one of these, embracing the women of his own clan, is prohibited to him; the other, to which women of the remaining three clans belong, is lawful.

Let us compare now the brother–sister relation in Melanesia and Europe. With us, the intimacy of childhood gradually cools off and changes into a somewhat constrained relation, in which the sister is naturally but not completely divided from her brother by social, psychological and biological factors. In Melanesia, as soon as any intimacy in play or in childish confidences might spring up, the strict taboo sets in. The sister remains a mysterious being, always near yet never intimate, divided by the invisible yet all powerful wall of traditional command which gradually changes into a moral and personal imperative. The sister remains the only spot on the sexual horizon permanently

hidden. Any natural impulses of infantile tenderness are as systematically repressed from the outset as other natural impulses
are in our children, and the sister becomes thus 'indecent' as an
object of thought, interest and feeling, just as the forbidden
things do for our children. Later on, as the personal experiences
in sexuality develop, the veil of reserve separating the two thick-
ens. Though they have constantly to avoid each other, yet, owing
to the fact that he is the provider of her household, they must
constantly keep one another in thought and attention. Such
artificial and premature repression must have its results. The
psychologists of the Freudian school could easily foretell them.

In all this I have spoken almost exclusively from the point of
view of the boy. What is the configuration of the Melanesian
girl's attitude to her family as it crystallizes at puberty? Roughly
speaking, her attitude does not differ so much from that of her
European counterpart as is the case with the boy. Just because of
the brother and sister taboo, the Trobriand matriarchy touches
the girl less than the boy. For, since her brother is strictly forbidden to take any interest in her sexual affairs, including her marriage, and her mother's brother has also to keep aloof from these
matters, it is, strangely enough, her father who is her guardian as
regards matrimonial arrangements. So that between father and
daughter not quite an identical, but a very similar relation exists
as with us. For among ourselves the friction between the female
child and her father is normally small, and thus the relation
approaches nearer to that found in the Trobriands between
father and child. There, on the other hand, the intimacy between
a grown-up man and an adolescent girl, who, be it remembered,
is not considered his kinswoman, is fraught with some temptation. This is not lessened but increased by the fact that though
the daughter is not actually tabooed by the laws of exogamy, yet
sexual intercourse between the two is considered in the highest
degree reprehensible, though it is never given the name of
suvasova, which means breach of exogamy. The reason for this

prohibition between father and daughter is, of course, simply that it is wrong to have sexual intercourse with the daughter of the woman with whom you co-habit. We shall not be astonished when later, as we trace the influence of the typical attitudes between members of the family, we shall find that father–daughter incest happens in reality, though it hardly could be called an obsession, nor has it any echo in folklore.

With regard to her mother, the general course of the relation is more natural than that in Europe, though not essentially different. One point of difference there is: namely, that the exodus of the girl at puberty from the parental home and her numerous outside sex interests normally prevent the development of mother–daughter rivalries and jealousies, though they do not always preclude the occurrence of father–daughter incest. Thus, with the exception of her attitude to the brother, broadly speaking, sentiments similar to those in Europe are to be found in a Melanesian girl.

9

THE COMPLEX OF MOTHER-RIGHT

We have been comparing the two civilizations, the European and the Melanesian, and we have seen that there exist deep differences, some of the forces by which society moulds man's biological nature being essentially dissimilar. Though in each there is a certain latitude given to sexual freedom, and a certain amount of interference with and regulation of the sex instinct, yet in each the incidence of the taboo and the play of sexual liberty within its prescribed bounds are entirely different. There is also a quite dissimilar distribution of authority within the family, and correlated with it a different mode of counting kinship. We have followed in both societies the growth of the average boy or girl under these divergent tribal laws and customs. We have found that at almost every step there are great differences due to the interplay between biological impulse and social rule which sometimes harmonize, sometimes conflict, sometimes lead to a short bliss, sometimes to an inequilibrium fraught, however, with possibilities for a future development. At the final stage of the child's life-history, after it has reached maturity, we have seen its feelings crystallize into a system of sentiments towards the mother, father, brother, sister, and in the Trobriands, the maternal uncle, a system which is typical of each society, and which, in order to adapt ourselves to psycho-analytic terminology, we called the 'Family Complex' or the 'nuclear complex'.

Now allow me to restate briefly the main features of these two 'complexes'. The Œdipus complex, the system of attitudes typical of our patriarchal society, is formed in early infancy, partly during the transition between the first and second stages of childhood, partly in the course of the latter. So that, towards its end, when the boy is about five or six years old, his attitudes are well formed, though perhaps not finally settled. And these attitudes comprise already a number of elements of hate and suppressed desire. In this, I think, our results do not differ to any extent from those of psycho-analysis.[1]

In the matrilineal society at that stage, though the child has developed very definite sentiments towards its father and mother, nothing suppressed, nothing negative, no frustrated desire forms a part of them. Whence arises this difference? As we saw, the social arrangements of the Trobriand matriliny are in almost complete harmony with the biological course of development, while the institution of father-right found in our society crosses and represses a number of natural impulses and inclinations. To trace it more in detail, there is the passionate attachment to the mother, the bodily desire to cling close to her, which in patriarchal institutions is in one way or another broken or interfered with; the influence of our morality, which condemns sexuality in children; the brutality of the father, especially

---

[1] I have come to realize since the above was written that no orthodox or semi-orthodox psycho-analyst would accept my statement of the 'complex', or of any aspect of the doctrine.

in the lower strata, the atmosphere of his exclusive right to mother and child acting subtly but strongly in the higher strata, the fear felt by the wife of displeasing her husband—all these influences force apart parents and children. Even where the rivalry between father and child for the mother's personal attention is reduced to a minimum, or to naught, there comes, in the second period, a distinct clash of social interests between father and child. The child is an encumbrance and an obstacle to the parental freedom, a reminder of age and decline and, if it is a son, often the menace of a future social rivalry. Thus, over and above the clash of sensuality, there is ample room for social friction between father and child. I say advisedly 'child' and not 'boy', for, according to our results, the sex difference between the children does not play any great part at this stage, nor has a closer relation between father and daughter as yet made its appearance.

All these forces and influences are absent from the matrilineal society of the Trobriands. First of all—and that has, bien entendu, nothing to do with matriliny—there is no condemnation of sex or of sensuality as such, above all, no moral horror at the idea of infantile sexuality. The sensuous clinging of the child to his mother is allowed to take its natural course till it plays itself out and is diverted by other bodily interests. The attitude of the father to the child during these two early periods is that of a near friend and helper. At the time when our father makes himself pleasant at best by his entire absence from the nursery, the Trobriand father is first a nurse and then a companion.

The development of pre-sexual life at this stage also differs in Europe and Melanesia; the repressions of the nursery among us, especially in the higher classes, develop a tendency towards clandestine inquisitions into indecent things, especially excretory functions and organs. Among the savages we find no such period. Now this infantile pre-genital indecency establishes distinctions between the decent–indecent, the pure–impure, and the indecent, parent-proof compartment reinforces and gives additional depth to the taboo which is suddenly cast over certain relations to the mother, that is to the premature banishment from her bed and bodily embraces.

So that here also the complications of our society are not shared by the children in the Trobriands. At the next stage of sexuality we find a no less relevant difference. In Europe there is a latency period more or less pronounced, which implies a breach of continuity in the sexual development and, according to Freud, serves to reinforce many of our repressions and the general amnesia, and to create many dangers in the normal development of sex. On the other hand, it also represents the triumph of other cultural and social interests over sexuality. Among the savages at this stage, sex in an early genital form—a form almost unknown among ourselves—establishes itself foremost among the child's interests, never to be dislodged again. This, while in many respects it is culturally destructive, helps the gradual and harmonious weaning of the child from the family influences.

With this we have entered already into the second half of the child's development, for the period of sexual latency in our society belongs to this part. When we consider these two later stages which form the second half of the development, we find another profound difference. With us during this early period of puberty, the Œdipus complex, the attitudes of the boy towards his parents, only solidify and crystallize. In Melanesia, on the other hand, it is mainly during this second epoch, in fact almost exclusively then, that any complex is formed. For only at this period is the child submitted to the system of repressions and taboos which begin to mould his nature. To these forces he responds, partly by adaptation, partly by developing more or less repressed antagonisms and desires, for human nature is not only malleable but also elastic.

The repressing and moulding forces in Melanesia are

twofold—the submission to matriarchal tribal law, and the prohibitions of exogamy. The first is brought about by the influence
of the mother's brother, who, in appealing to the child's sense of
honour, pride and ambition, comes to stand to him in a relation
in many respects analogous to that of the father among us. On
the other hand, both the efforts which he demands and the
rivalry between successor and succeeded introduce the negative
elements of jealousy and resentment. Thus an 'ambivalent' attitude is formed in which veneration assumes the acknowledged
dominant place, while a repressed hatred manifests itself only
indirectly.

The second taboo, the prohibition of incest, surrounds the
sister, and to a lesser degree other female relatives on the maternal side, as well as clanswomen, with a veil of sexual mystery. Of
all this class of women, the sister is the representative to whom
the taboo applies most stringently. We noted that this severing
taboo, entering the boy's life in infancy, cuts short the incipient
tenderness towards his sister which is the natural impulse of a
child. This taboo also, since it makes even an accidental contact
in sexual matters a crime, causes the thought of the sister to be
always present, as well as consistently repressed.

Comparing the two systems of family attitudes briefly, we see
that in a patriarchal society, the infantile rivalries and the later
social functions introduce into the attitude of father and son,
besides mutual attachment, also a certain amount of resentment
and dislike. Between mother and son, on the other hand, the
premature separation in infancy leaves a deep, unsatisfied craving which, later on, when sexual interests come in, is mixed up
in memory with the new bodily longings, and assumes often an
erotic character which comes up in dreams and other fantasies.
In the Trobriands there is no friction between father and son,
and all the infantile craving of the child for its mother is allowed
gradually to spend itself in a natural, spontaneous manner. The
ambivalent attitude of veneration and dislike is felt between a

man and his mother's brother, while the repressed sexual attitude of incestuous temptation can be formed only towards his sister. Applying to each society a terse, though somewhat crude formula, we might say that in the Œdipus complex there is the repressed desire to kill the father and marry the mother, while in the matrilineal society of the Trobriands the wish is to marry the sister and to kill the maternal uncle.

With this, we have summarized the results of our detailed inquiry, and given an answer to the first problem set out at the beginning, that is, we have studied the variation of the nuclear complex with the constitution of the family, and we have shown in what manner the complex depends upon some of the features of family life and sexual morals.

We are indebted to psycho-analysis for the discovery that there exists a typical configuration of sentiments in our society, and for a partial explanation, mainly concerned with sex, as to why such a complex must exist. In the foregoing pages we were able to give an outline of the nuclear complex of another society, a matrilineal one, where it has never been studied before. We found that this complex differs essentially from the patriarchal one, and we have shown why it must differ and what social forces bring it about. We have drawn our comparison on the broadest basis, and, without neglecting sexual factors, we have also systematically drawn in the other elements. The result is important, for, so far, it has never been suspected that another type of nuclear complex might be in existence. By my analysis, I have established that Freud's theories not only roughly correspond to human psychology, but that they follow closely the modification in human nature brought about by various constitutions of society. In other words, I have established a deep correlation between the type of society and the nuclear complex found there. While this is in a sense a confirmation of the main tenet of Freudian psychology, it might compel us to modify certain of its features, or rather to make some of its formulae more elastic. To put it concretely, it appears necessary to draw in more systematically the correlation between biological and social influences; not to assume the universal existence of the Œdipus complex, but in studying every type of civilization, to establish the special complex which pertains to it.

Part II

The Mirror of Tradition

1

COMPLEX AND MYTH
IN MOTHER-RIGHT

It now remains to proceed to the study of the second problem posed in the first part of this volume; that is to investigate whether the matrilineal complex, so entirely different in its genesis and its character from the Œdipus complex, exercises also a different influence on tradition and social organization; and to show that in the social life, as well as in the folklore, of these natives their specific repressions manifest themselves unmistakably. Whenever the passions, kept normally within traditional bounds by rigid taboos, customs and legal penalties, break through in crime, perversion, aberration, or in one of those dramatic occurrences which shake from time to time the humdrum life of a savage community—then these passions reveal the matriarchal hatred of the maternal uncle or the incestuous wishes towards the sister. The folklore of these Melanesians also mirrors the matrilineal complex. The examination of myth, fairy tales and legend, as well as of magic, will show that the repressed hatred of the maternal uncle, ordinarily masked by conventional reverence and solidarity, breaks through in those narratives constructed on the model of the day-dream and dictated by repressed longings.

Especially interesting is the magic of love of these natives and the mythology connected with it. All sexual attraction, all power of seduction is believed to reside in the magic of love. This magic again the natives regard as founded in a dramatic occurrence of the past, narrated in a strange tragic myth of brother and sister incest. Thus the position established by the description of social relations within the family, an analysis of kinship, can also be independently demonstrated by the study of the culture of these Melanesian natives.

2

DISEASE AND PERVERSION

The evidence adduced in this part of the essay is not quite homogeneous, while on some points I have had full information, I shall have to confess my ignorance or only incomplete knowledge in others, and there I shall indicate the problem rather than solve it. This is due partly to my lack of expert knowledge of mental disease, partly to my having found it impossible to psycho-analyze the natives by the orthodox technique; partly to an unavoidable unevenness in the material, especially that which I collected among other tribes where I resided for a much shorter time and worked under less favourable conditions than in the Trobriands.

I shall start with the weakest items in my répertoire. Here comes first the question of neurosis and mental disease. We have seen in the comparative account of the child's development among ourselves and in the Trobriands that the matrilineal complex is formed later in the life of a child, that it is formed outside the intimacy of the family circle, that it entails fewer shocks, if any, that it is due mainly to the play of rivalry, while its erotic thwartings do not go to the roots of infantile sexuality. Since this is so, the Freudian theory of neurosis would lead us to expect a much smaller prevalence of those neuroses (*Übertragungsneurosen*) due to the traumas of childhood. It is a great pity that a competent alienist has not been able to examine the Trobrianders under the same conditions as myself, for I think he could throw some interesting sidelights upon the assumptions of psycho-analysis.

When studying the Trobrianders, it would be futile for an ethnographer to compare them with Europeans, for with us there are innumerable other factors which complicate the picture and contribute to the formation of mental disease. But some thirty miles south of the Trobriands there are the Amphlett Islands, inhabited by people essentially similar in race, custom, and language, but who differ, however, very much in social organization, have strict sexual morals, that is, regard pre-nuptial sexual intercourse with disapproval and have no institutions to support sexual license, while their family life is much more closely knit. Though matrilineal, they have a much more developed patriarchal authority, and this, combined with the sexual repressiveness, establishes a picture of childhood more similar to our own.[1]

Now even with my own limited knowledge of the subject, I received quite a different impression of the neurotic dispositions of these natives. In the Trobriands, though I knew scores of natives intimately and had a nodding acquaintance with many more, I could not name a single man or woman who was hysterical or even neurasthenic. Nervous tics, compulsory actions or obsessive ideas were not to be found. In the system of native pathology, based, of course, on belief in black magic, but reasonably true to the symptoms of disease, there are two categories

[1] For a description of some customs and features of the culture of the natives of the Amphlett Island. see the author's *Argonauts of the Western Pacific*, chap. xi.

of mental disorder—*nagowa*, which corresponds to cretinism, idiocy, and is also given to people who have a defect of speech; and *gwayluwa*, which corresponds roughly to mania, and comprises those who from time to time break out into acts of violence and deranged behaviour. The natives of the Trobriands know well and recognize that in the neighbouring islands of the Amphletts and d'Entrecasteaux there are other types of black magic which can produce effects on the mind different from those known to themselves, of which the symptoms are according to their accounts compulsory actions, nervous tics and various forms of obsession. And during my few months' stay in the Amphletts, my first and strongest impression was that this was a community of neurasthenics. Coming from the open, gay, hearty and accessible Trobrianders, it was astonishing to find oneself among a community of people distrustful of the newcomer, impatient in work, arrogant in their claims, though easily cowed and extremely nervous when tackled more energetically. The women ran away as I landed in their villages and kept in hiding the whole of my stay, with the exception of a few old hags. Apart from this general picture, I at once found a number of people affected with nervousness whom I could not use as informants, because they would either lie in some sort of fear, or else become excited and offended over any more detailed questioning. It is characteristic that in the Trobriands even the spiritualistic mediums are *poseurs* rather than abnormal people. And while in the Trobriands black magic is practised in a 'scientific' manner by men, that is by methods which present small claim to the supernatural, in the islands of the south there are 'flying wizards' who practise the magic which in other parts belongs only to semi-fabulous witches, and who make at first sight a quite abnormal impression.

In another community among whom I served my ethnographic apprenticeship, and whom I therefore did not study with the same methods or come to know as intimately as I did the Trobrianders, the conditions are even more repressive than in the Amphlett Islands. The Mailu, inhabiting a portion of the south coast of New Guinea, are patrilineal, have a pronounced paternal authority in the family, and a fairly strict code of repressive sexual morals.[1] Among these natives, I had noted a number of people whom I had classed as neurasthenics, and therefore useless as ethnographic informants.

But all these tentative remarks, though they are not sheer guesses, are intended only to raise the problem, and to indicate what the solution would most probably be. The problem would therefore be: to study a number of matrilineal and patriarchal communities of the same level of culture, to register the variation of sexual repression and of the family constitution, and to note the correlation between the amount of sexual and family repression and the prevalence of hysteria and compulsion neurosis. The conditions in Melanesia, where side by side we find communities living under entirely different conditions, are like a naturally arranged experiment for this purpose.

Another point which might be interpreted in favour of the Freudian solution of this problem is the correlation of sexual perversions with sexual repression. Freud has shown that there is a deep connection between the course of infantile sexuality and the occurrence of perversion in later life. On the basis of his theory, an entirely lax community like that of the Trobrianders, who do not interfere with the free development of infantile sexuality, should show a minimum of perversions. This is fully confirmed in the Trobriands. Homosexuality was known to exist in other tribes and regarded as a filthy and ridiculous practice. It cropped up in the Trobriands only with the influence of white

[1] Compare the writer's monograph on 'The Natives of Mailu' in the *Proceedings of the Royal Society of Australia*, vol. 39, 1915. No information on mental disease is contained there. I had hoped to return to the district and the essay was published as a preliminary account in which I did not include all I knew and had noted, thinking of republishing it in a fuller form.

man, more especially of white man's morality. The boys and girls on a Mission Station penned in separate and strictly isolated houses, cooped up together, had to help themselves out as best they could, since that which every Trobriander looks upon as his due and right was denied to them. According to very careful inquiries made of non-missionary as well as missionary natives, homosexuality is the rule among those upon whom white man's morality has been forced in such an irrational and unscientific manner. At any rate, there were a few cases in which 'evil doers' caught in *flagrante delicto*, were ignominiously banished from the face of God back into the villages, where one of them tried to continue it, but had to give up under the pressure of the native morals, expressed in scorn and derision. I have also reason to suppose that perversions are much more prevalent in the Amphlett and d'Entrecasteaux archipelago in the south, but again I have to regret that I was not able to study this important subject in detail.

3

DREAMS AND DEEDS

Now we have to study how the integral sentiment of the matrilineal family in the Trobriands expresses itself in the culture and
social organization of the natives. If we pushed this problem too
deep it would indeed lead us to a minute examination from this
point of view of practically every manifestation of their tribal
life. We shall have to make a selection and pick out the most
relevant domains of fact. These can be divided into two categories: (1) the free fantasies, and (2) the data of folklore. To the first
class belong those products of individual imagination such as
dreams, day-dreams, personal desires and ideals which, coming
from the individual's own life, are shaped by the endo-psychic
forces of his personality. In this class can be reckoned not only
the manifestations of fantasies in thought and dream, but also
in deed. For a crime or a sin or an act which outrages public
opinion and decency is committed when the repressive forces of
law and morality are broken by the repressed passions. In such
deeds we can measure both the strength of the ideal and the
depth of the passion. We shall turn now to this first class of

dreams and deeds in which the individual shakes off temporarily the shackles of custom and reveals the repressed elements and the conflict with the repressing forces.

Dreams and day-dreams are not an easy subject for study among the Melanesians of the Trobriand Islands. It is a remarkable and characteristic feature of these natives, in which they seem to differ from other savages, that they apparently dream little, have little interest in their dreams, seldom relate them spontaneously, do not regard the ordinary dream as having any prophetic or other importance, and have no code of symbolic explanation whatever. When I tackled the subject directly, as I often did, and asked my informants whether they had dreamt, and, if so, what their dreams had been, the answer was usually negative, with rare exceptions, to which we will return. Is this absence of dreams, or rather of interest in dreams, due to the fact that we are dealing with a non-repressed society, a society among whom sex as such is in no way restricted? Is it so because their 'complex' is weak, appears late, and has few infantile elements? This rarity of free dreams and the absence of strong effect, hence absence of remembrance, point to the same conclusion as the absence of neurosis, that is, to the correctness in broad outline of the Freudian theory. For this theory affirms that the main cause of dreams is unsatisfied sexual appetite, and especially such sexual or quasi-sexual impulses as are repressed violently in infancy. To this question one could only obtain a satisfactory answer by collecting rich comparative material among two communities of similar culture and mode of living but with different repressions.

I have used so far the expression 'free dreams', for there is a class of dreams which it is difficult to range, whether with the free or with the fixed fantasies, since they run on lines prescribed by tradition and could be called 'official dreams'. Such, for instance, are dreams in which a man leading an enterprise or carrying out some task is supposed to dream under certain circumstances about the object of his enterprise. The leaders of fishing excursions dream about the weather, about the place where the shoals may appear, about the best date for the expedition, and they give their orders and instructions accordingly. Those in charge of the overseas expeditions called Kula are often supposed to have dreams about the success of their ceremonial trading. Above all the magicians have dreams associated with the performance of their magic. There is also another form of typical or traditional dream associated with magic, that, namely, which comes about as the direct result of a spell or of a rite. Thus, in the ceremonial overseas trading there is a certain spell which acts directly on the mind of the partner, induces in him a dream, and this dream makes the partner desire the exchange. Most love magic is supposed to produce a dream which awakens the amorous wish. Thus these natives, remarkably enough, reverse the Freudian theory of dreams, for to them the dream is the cause of the wish.[1] In reality, this class of traditional dreams is very much within the lines of the Freudian theory. For they are constructed as a projection on to the victim of the magician's own desire. The victim of love magic feels in her dream an itching, a craving which is the same as the state of mind of the performer of the magic. The Kula partner under the influence of magic is supposed to dream of glorious scenes of exchange which form the very vision dominating the wishes of the performer.

Nor are such dreams merely spoken of and only supposed to exist. Very frequently the magician himself would come and tell me that he had dreamt about a good yield in fishing, and would organize an expedition on the strength of it. Or a garden wizard would speak of a dream he had had about a long drought, and therefore order certain things to be done. During the annual

---

[1] Cf. also my *Argonauts of the Western Pacific*, chapter on magic and detailed descriptions of the rites and spells in the course of the narrative.

ceremonial feast in honour of dead ancestors I had on two occasions opportunities of noting dreams of natives. In both cases the dream referred to the proceedings, and in one the dreamer claimed to have dreamt that he had had a conversation with the spirits, who were not satisfied with things. Another class of typical dream is concerned with the birth of babies. In these the future mother has a sort of dream annunciation from one of her dead relatives.[1]

Now one of the typical or official dreams is the sexual dream, which interests us here more especially. A man will dream that a woman visits him at night; in the dream he will have congress with her, and he will awake finding the discharge of semen on the mat. This he will conceal from his wife, but he will try to follow up the dream actively in real life and initiate an intrigue with the woman. For this dream means that she who visited him had performed love magic and that she desires him.

About such dreams I had a number of personal confidences, followed by the story of the subsequent efforts of the man at establishing an intrigue with his dream visitor.

Now, naturally, as soon as I was told by the natives about their erotic dreams, I was at once keenly on the scent of incestuous dreams. To the question: 'Do you ever dream of your mother in this way?' the answer would be a calm, unshocked negation. 'The mother is forbidden—only a *tonagowa* (imbecile) would dream such a thing. She is an old woman. No such thing would happen.' But whenever the question would be put about the sister, the answer would be quite different, with a strong affective reaction. Of course I knew enough never to ask such a question directly of a man, and never to discuss it in company. But even asking in the form of whether 'other people' could ever have such dreams, the reaction would be that of indignation and anger. Sometimes there would be no answer at all; after an

---

[1] Cf. 'Baloma'—article in the *Journal of the R. Anthrop. Inst.*, 1916.

embarrassed pause another subject would be taken up by the informant. Some, again, would deny it seriously, others vehemently and angrily. But, working out the question bit by bit with my best informants, the truth at last appeared, and I found that the real state of opinion is different. It is actually well known that 'other people' have such dreams—'a man is sometimes sad, ashamed, and ill-tempered. Why? Because he has dreamt that he had connection with his sister.' 'This made me feel ashamed,' such a man would say. I found that this is, in fact, one of the typical dreams known to exist, occurring frequently, and one which haunts and disturbs the dreamer. That this is so, we will find confirmed by other data, especially in myth and legend.

Again, the brother–sister incest is the most reprehensible form of breach of the rules of exogamy—which institution makes it illicit to have connection with any woman of the same clan. But though the brother–sister incest is regarded with the utmost horror, a breach of clan exogamy is a thing both smart and desirable, owing to the piquant difficulties in carrying it out. In accordance with this, dreams about clan incest are very frequent. Thus, comparing the different types of incestuous dreams, there is every reason to assume that the mother hardly ever appears in them and, if she does, these dreams leave no deep impression; that the more distant female relatives are dreamt of frequently, and that the impression left is pleasant; while incestuous dreams about the sister occur and leave a deep and painful memory. This is what might have been expected, for, as we saw when following the development of their sexuality, there is no temptation in the case of the mother, a violent and strongly repressed one towards the sister, and a spicy, not very repressed prohibition about the clanswoman.

Brother and sister incest is regarded with such horror by the natives that at first an observer, even well acquainted with their life, would confidently affirm that it would never occur, though a Freudian might have his suspicions. And these, on closer search,

would be found fully justified. Incest between brother and sister
existed even in olden days, and there are certain family scandals
told especially about the ruling clan of the Malasi. Nowadays,
when the ancient morals and institutions break down under the
influence of spurious Christian morality and the white man's so-
called law and order is introduced, the passions repressed by
tribal tradition break through even more violently and openly. I
have three or four cases on record in which public opinion
definitely, though in whispered undertones, accused a brother
of incestuous relations with his sister. One case, however, stands
out, for it was a lasting intrigue famous for its effrontery, for the
notorious character of the hero and heroine, and for the
scandalous stories spun around it.

Mokadayu, of Okopukopu, was a famous singer. Like all of his
profession, he was no less renowned for his success with ladies.
'For,' say the natives, 'the throat is a long passage like the *wila*
(vagina), and the two attract each other.' 'A man who has a
beautiful voice will like women very much and they will like
him.' Many stories are told of how he slept with all the wives of
the chief in Olivilevi, how he seduced this and that married
woman. For a time, Mokadayu had a brilliant and very lucrative
career as a spiritualistic medium, extraordinary phenomena
happening in his hut, especially dematerializations of various
valuable objects thus transported to the spirit land. But he was
unmasked, and it was proved that the dematerialized objects had
merely remained in his own possession.

Then there came about the dramatic incident of his incestuous
love with his sister. She was a very beautiful girl, and, being a
Trobriander, she had, of course, many lovers. Suddenly she with-
drew all her favours and became chaste. The youth of the village,
who confided to each other their banishment from her favours,
decided to find out what was the matter. It soon appeared that,
whoever might be the privileged rival, the scene must be laid in
her parental house. One evening when both parents were away, a

hole was made in the thatch and through it the discarded lovers saw a sight which shocked them deeply; brother and sister were caught in *flagrante delicto*. A dreadful scandal broke out in the village, which, in olden days, would certainly have ended in the suicide of the guilty pair. Under the present conditions they were able to brave it out and lived in incest for several months till she married and left the village.

Besides the actual brother and sister incest, there is, as I have said, a breach of exogamous rules which is called *suvasova*. A woman of the same clan is forbidden to a man under the penalty of shame and an eruption of boils all over the body. Against this second ailment there is a magic, which, as many of my informants told me with a self-satisfied smirk, is absolutely efficacious. The moral shame of such incidents is in reality small, and as with many other rules of official morality, he who breaks it is a smart fellow. A young man who is a real Don Juan, and who has a good conceit of himself, will scorn the unmarried girls, and try always to have an intrigue with a married woman, especially a chief's wife, or else commit acts of *suvasova*. The expression '*suvasova yoku*', 'Oh, thou exogamy breaker!' sounds something like, 'Oh, you gay dog!' and is a facetious compliment.

To complete the picture, the negative evidence may be stated here that not one single case of mother–son incest could be found, not even a suspicion of it, though the loudness and stringency of the taboo is by no means so great as in the brother–sister incest. In the summary given above of the typical family sentiments among the Trobrianders, I have stated that the relations between father and daughter are the only ones built up on the same pattern as in a patriarchal society. As could be expected therefore, father–daughter incest is of by no means rare occurrence. Two or three cases in which there seem to be no doubt whatever are on record. One of them concerned a girl, who, besides her relations with her father, was the sweetheart of a local boy then in my service. He wanted to marry her, and

appealed to me for financial and moral support in this enterprise; I therefore had full details of the incest, which left me in
no doubt whatever about the relationship and its long duration.

So far we have spoken about the sexual taboo and the
repressed wish to break it, which finds expression in dreams, in
acts of crime and passion. There is, however, another relation
fraught with repressed criminal desires, that of a man towards
his matriarch, the brother of his mother. With regard to dreams,
there is one interesting fact to be noted here: the belief, namely,
that in prophetic dreams of death it will always be a *veyola* (real
kinsman), usually the sister's son, who will foredream his
uncle's death. Another important fact belonging to the sphere of
action and not of dreams is connected with witchcraft. A man
who has acquired the black magic of disease must choose his
first victim from among his near maternal relatives. Very often a
man is said to choose his own mother. So that when anyone is
known to be learning sorcery, his real kinsmen, that means his
maternal relatives, are always frightened and on the look out for
personal danger.

In the chronicles of actual crime, there are also several cases to
be registered, bearing on our problem. One of them happened
in the village of Osapola, half an hour away from where I lived at
that time, and I knew the actors well. There were three brothers,
the eldest blind. The youngest one used always to take the betel
nut before it properly ripened and deprive the blind man of his
share. The blind man one day got in a dreadful fury and, seizing
an axe, somehow managed to wound the youngest brother. The
middle one then took a spear and killed the blind one. He was
sentenced to twelve months' imprisonment by the white resi-
dent magistrate. The natives regarded this as an outrageous
injustice. The killing of one brother by another is a purely
internal matter, certainly a dreadful crime and an awful tragedy,
but one with which the outer world is in no way concerned, and
it can only stand by and show its horror and pity. There are other

cases of violent quarrels, fights, and one or two more murders within the matrilineal family, which I have on record.

Of parricide, on the other hand, there is not one single case to be cited. Yet to the natives, as I have said, parricide would be no special tragedy, and would be merely a matter to be settled with the father's own clan.

Apart from the dramatic events, the crimes and tragedies which shake the tribal order to its very foundations, there are the small events which indicate merely the boiling of the passions under the apparently firm and quiet surface. For, as we saw, society builds up its traditional norms and ideals, and sets up trammels and barriers to safeguard them. Yet these very trammels provoke certain emotional reactions.

Nothing surprised me so much in the course of my socio-logical researches as the gradual perception of an undercurrent of desire and inclination running counter to the trend of conven-tion, law and morals. Mother-right, the principle that unity of kinship exists only in the mother line, and that this unity of kinship should claim all affection, as well as all duties and loyal-ties, is the dictate of tradition. But in reality friendship and affec-tion to the father, community of personal interest and desires with him, combined with the wish to shake off the exogamous trammels of the clan—these are the live forces which flow from personal inclination and the experiences of individual life. And these forces contribute much to fan ever-present sparks of enmity between brothers, and between the mother's brother and the nephew. So that in the real feelings of the individual, we have, so to speak, a sociological negative of the traditional principle of matriliny.[1]

---

[1] This point has been elaborated by the writer in *Crime and Custom*, 1926.

4

OBSCENITY AND MYTH

We now proceed to the discussion of folklore in relation to the typical sentiments of the matrilineal family, and with this we enter the best cultivated plot on the boundary of psycho-analysis and anthropology. It has long been recognized that for one reason or another the stories related seriously about ancestral times and the narratives told for amusement correspond to the desires of those among whom they are current. The school of Freud maintain, moreover, that folklore is especially concerned with the satisfaction of repressed wishes by means of fairy tales and legends; and that this is the case also with proverbs, typical jokes and sayings and stereotyped modes of abuse.

Let us begin with these last. Their relation to the unconscious must not be interpreted in the sense that they satisfy the repressed cravings of the person abused, or even of the abuser. For instance, the expression widely current among oriental races and many savages, 'eat excrement', as well as in a slightly modified form among the Latins, satisfies directly the wish of neither. Indirectly it is only meant to debase and disgust the person thus addressed. Every form of abuse or bad language contains certain propositions fraught with strong emotional possibilities. Some bring into play emotions of disgust and shame; others again draw attention to, or impute, certain actions which are considered abominable in a given society, and thus wound the feelings of the listener. Here belongs blasphemy, which in European culture reaches its zenith of perfection and complexity in the innumerable variations of '*Me cago en Dios!*' pullulating wherever the sonorous Spanish is spoken. Here, also, belong all the various abuses by reference to social position, despised or degraded occupations, criminal habits, and the like, all of them very interesting sociologically, for they indicate what is considered the lowest depth of degradation in that culture.

The incestuous type of swearing, in which the person addressed is invited to have connection with a forbidden relative, usually the mother, is in Europe the speciality of the Slavonic nations, among whom the Russians easily take the lead, with the numerous combinations of 'Yob twayn mat' ('Have connection with thy mother'). This type of swearing interests us most, because of its subject, and because it plays an important part in the Trobriands. The natives there have three incestuous expressions: '*Kwoy inam*'—' Cohabit with thy mother'; '*Kwoy lumuta*'—'cohabit with thy sister'; and '*Kwoy um' kwava*'—'cohabit with thy wife.' The combination of the three sayings is curious in itself, for we see, side by side, the most lawful and the most illicit types of intercourse used for the same purpose of offending and hurting. The gradation of intensity is still more remarkable. For while the invitation to maternal incest is but a mild term used in chaff or as a joke, as we might say, 'Oh, go to Jericho', the mention of sister incest in abuse is a most serious offence, and one used only when real anger is aroused. But the worst insult, one which I have known to be seriously used at the most twice, and once, indeed, it was among the causes of the incident of fratricide described above, is the imperative to have connection

with the wife. This expression is so bad that I learnt of its existence only after a long sojourn in the Trobriands, and no native
would pronounce it but in whispers, or consent to make any
jokes about that incongruous mode of abuse.

What is the psychology of this gradation? It is obvious that it
stands in no distinct relation to the enormity or unpleasantness
of the act. The maternal incest is absolutely and completely out
of the question, yet it is the mildest abuse. Nor can the criminal-
ity of the action be the reason for the various strengths of the
swearing, for the least criminal, in fact the lawful connection, is
the most offensive when imputed. The real cause is the plausibility and the reality of the act, and the feeling of shame, anger, and
social degradation at the barriers of etiquette being pulled down
and the naked reality brought to light. For the sexual intimacy
between husband and wife is masked by a most rigid etiquette,
not so strict of course as that between brother and sister, but
directly aiming at the elimination of any suggestive modes of
behaviour. Sexual jokes and indecencies must not be pronounced
in the company of the two consorts. And to drag out the personal, direct sexuality of the relation in coarse language is a
mortal offence to the sensitiveness of the Trobrianders. This
psychology is extremely interesting, just because it discloses that
one of the main forces of abuse lies in the relation between the
reality and plausibility of a desire or action and its conventional
repressions.

The relation between the abuse by mother and by sister incest
is made clear by the same psychology. Its strength is measured
mainly by the likelihood of reality corresponding to the imput-
ation. The idea of mother incest is as repugnant to the native as
sister incest, probably even more. But just because, as we saw, the
whole development of the relationship and of sexual life makes
incestuous temptations of the mother almost absent, while the
taboo against the sister is imposed with great brutality and kept
up with rigid strength, the real inclination to break the strong

taboo is much more actual. Hence this abuse wounds to the quick.

There is nothing to be said about proverbs in the Trobriands, for they do not exist. As to the typical sayings and other linguistic uses, I shall mention here the important fact of the word *luguta*, my sister, being used in magic as a word which signifies incompatibility and mutual repulsion.

We pass now to myth and legend, that is, to the stories told with a serious purpose in explanation of things, institutions, and customs. To make the survey of this very extensive and rich material clear yet rapid, we shall classify these stories into three categories: (1) Myths of the origin of man, and of the general order of society, and especially totemic divisions and social ranks; (2) Myths of cultural change and achievements which contain stories about heroic deeds, about the establishment of customs, cultural features and social institutions; (3) Myths associated with definite forms of magic.[1]

The matrilineal character of the culture meets us at once in the first class, that is, in the myths about the origins of man, of the social order, especially chieftainship and totemic divisions, and of the various clans and sub-clans. These myths, which are numerous, for every locality has its own legends or variations, form a sort of connected cycle. They all agree that human beings have emerged from underground through holes in the earth. Every sub-clan has its own place of emergence, and the events which happened on this momentous occasion determined sometimes the privileges or disabilities of the sub-clan. What interests us most in them is that the first ancestral groups whose appearance is mentioned in the myth consist always of a woman, sometimes accompanied by her brother, sometimes by the totemic animal, but never by a husband. In some of the myths

---

[1] Cf. the chapter on Mythology in *Argonauts of the Western Pacific*, especially pp. 304 sqq.

the mode of propagating of the first ancestress is explicitly described. She starts the line of her descendants by imprudent exposure to the rain or, lying in a grotto, is pierced by the dripping of the stalactites; or bathing she is bitten by a fish. She is 'opened up' in this way, and a spirit child enters her womb and she becomes pregnant.[1] Thus instead of the creative force of a father, the myths reveal the spontaneous procreative powers of the ancestral mother.

Nor is there any other rôle in which the father appears. In fact, he is never mentioned, and does not exist in any part of the mythological world. Most of these local myths have come down in very rudimentary form, some containing only one incident or an affirmation of right and privilege. Those of them which contain a conflict or a dramatic incident, elements essential in ungarbled myth, depict invariably a matrilineal family and the drama happening within it. There is a quarrel beteen two brothers which makes them separate, each taking his sister. Or, again, in another myth, two sisters set out, disagree, separate and found two different communities.

In a myth which might perhaps be classed in this group, and which accounts for the loss of immortality, or, to put it more correctly, of perpetual youth by human beings, it is the quarrel between grandmother and granddaughter which brings about the catastrophe. Matriliny—in the fact that descent is reckoned by the female—mother-right—in the great importance of the

---

[1] Freudians will be interested in the psychology of symbolism underlying these myths. It must be noted that the natives have no idea whatever of the fertilizing influence of the male semen, but they know that a virgin cannot conceive, and that to become a mother a woman has to be 'opened up' as they express it. This in the everyday life of the village is done at an early age by the appropriate organ. In the myth of the primeval ancestress, where the husband or any sexually eligible male companion is excluded, some natural object is selected, such as a fish or a stalactite. Cf. for further material on this subject my article in *Psych*, Oct., 1923, reprinted as *The Father in Primitive Psychology*, 1927.

part played by women, the matriarchal configuration of kinship, in the dissensions of brothers—in short, the pattern of the matrilineal family, is evident in the structure of myths of this category. There is not a single myth of origins in which a husband or a father plays any part, or even makes his appearance. That the matrilineal nature of the mythological drama is closely associated with the matrilineal repressions within the family should need no further argument to convince a psycho-analyst.

Let us now turn to the second class of myths, those referring to certain big cultural achievements brought about by heroic deeds and important adventures. This class of myth is less rudimentary, consists of long cycles, and develops pronouncedly dramatic incidents. The most important cycle of this category is the myth of Tudava, a hero born of a virgin who was pierced by the action of stalactite water. The deeds of this hero are celebrated in a number of myths, which differ slightly according to the district in which they are found, and which ascribe to him the introduction of agriculture and the institution of a number of customs and moral rules, though his own moral character is very weakly developed. The main deed of this hero, however, the one known all over the district, and forming the bedrock of all the myths, is the slaying of an ogre. The story runs as follows:

Humanity led a happy existence in the Trobriand Archipelago. Suddenly a dreadful ogre called Dokonikan made his appearance in the eastern part of the islands. He fed on human flesh and gradually consumed one community after another. At the northwestern end of the island in the village of Laba'i there lived at that time a family consisting of a sister and her brothers. When Dokonikan ranged nearer and nearer to Laba'i the family decided to fly. The sister, however, at that moment wounded her foot and was unable to move. She was therefore abandoned by her brothers, who left her with her little son in a grotto on the beach of Laba'i, and sailed away in a canoe to the south-west. The boy was brought up by his mother, who taught him first the choice

of proper wood for a strong spear, then instructed him in the
Kwoygapani magic which steals away a man's understanding. The
hero sallied forth, and after having bewitched Dokonikan with
the *Kwoygapani* magic, killed him and cut off his head. After that
he and his mother prepared a taro pudding, in which they hid
and baked the head of the ogre. With this gruesome dish Tudava
sailed away in search of his mother's brother. When he found
him he gave him the pudding, in which the uncle with horror
and dismay found the head of Dokonikan. Seized with fear and
remorse, the mother's brother offered his nephew all sorts of
gifts in atonement for having abandoned him and his mother to
the ogre. The hero refused everything, and was only appeased
after he had received his uncle's daughter in marriage. After that
he set out again and performed a number of cultural deeds,
which do not interest us further in this context.

In this myth there are two conflicts which set the drama in
motion: first the cannibalistic appetite of the ogre, and second
the abandonment of mother and son by the maternal uncle. The
second is a typical matrilineal drama, and corresponds distinctly
to the natural tendency, repressed by tribal morals and custom,
as we have found it in our analysis of the matrilineal family in
the Trobriands. For the mother's brother is the appointed guardian of her and her family. Yet this is a duty which both weighs
heavily upon him, and is not always gratefully and pleasantly
received by his wards. Thus it is characteristic that the opening
of the most important heroic drama in mythology should be
associated with a capital sin of the matriarch's neglect of his
duty.

But this second matriarchal conflict is not altogether
independent of the first. When Dokonikan is killed his head is
presented in a dish of wood to the maternal uncle. If it were only
to frighten him by the sight of the monster, there would be no
point in disguising the head in the taro pudding. Moreover, since
Dokonikan was the general enemy of humanity, the sight of his

head should have filled the uncle with joy. The whole setting of this incident and the emotion which underlies it, receive meaning only if we assume that there is some sort of association or connivance between the ogre and the uncle. In that case, to give one cannibal's head to be eaten by the other is just the right sort of punishment, and the story contains then in reality one villain and one conflict distributed over two stages and duplicated into two persons. Thus we see that the legend of Tudava contains a typical matrilineal drama which forms its core, and which is brought to a logical conclusion. I shall remain satisfied, therefore, with having pointed out those features which are indisputable, and are clearly contained in the facts themselves, and I shall not enter in detail into further interpretations of this myth, which would necessitate certain historical and mythological hypotheses. But I wish to suggest that the figure of Dokonikan is not altogether explained by his association with the matriarch, that he may be a figure handed from a patriarchal culture into a matriarchal one, in which case he might represent the father and husband. If this be so, the present legend would be extremely interesting in showing how the prevalent cast of a culture moulds and transforms persons and situations to fit them into its own sociological context.

Another incident in this myth which I shall only indicate here, is the marriage at the end of the story of the hero to his maternal cross-cousin. This, in the present kinship system of the natives, is considered distinctly an improper thing, though not actually incestuous.

Passing to another legendary cycle, we have the story of two brothers who quarrel over a garden plot—as so often happens in real life—and in this quarrel the elder kills the younger. The myth does not relate any compunction for this act. It describes, instead, in detail the culinary anti-climax of the drama; the elder brother digs a hole in the ground, brings stones, leaves and firewood, and, as if he had just killed a pig or hauled out a big fish, he proceeds to bake his brother in an earthen oven. Then he hawks the baked flesh about from one village to another, rebaking it from time to time when his olfactory sense indicates the necessity of such a procedure. Those communities which decline his offer remain non-cannibalistic; those which accept become flesh-eaters ever afterwards. Thus here cannibalism is traced to a fratricidal act, and to preference or dislike for a food thus criminally and sinfully obtained. Needless to say, this is the myth of the non-cannibalistic tribes only. The same difference between cannibalism and its absence is explained by the man-eating natives of Dobu and the other cannibalistic districts of the d'Entrecasteaux Islands by a story in which cannibalism is certainly not branded as anything unpleasant. This story also, however, consists in a difference, if not in an actual quarrel between two brothers and two sisters.[1] What mainly interests us in these myths is the matrilineal imprint which they possess in the quarrel between elder and younger brother.

The myth about the origins of fire, which also contains a brief mention of the origins of sun and moon, describes dissension between two sisters. It may be added that fire in this myth is described as originating in a woman's sexual organs.

The reader accustomed to psycho-analytic interpretations of myth and to psychological and anthropological writings on the subject in general, will find all my remarks singularly simple and unsophisticated. All that is said here is clearly written on the surface of the myth, and I have hardly attempted any complicated or symbolic interpretation. This, however, I refrained from doing on purpose. For the thesis here developed that in a matriarchal society myth will contain conflicts of a specifically matrilineal nature is better served if supported only by unquestionable arguments. Moreover, if I am right, and if our sociological point

[1] These myths have already been given in *Argonauts of the Western Pacific*, Chapter on 'Mythology', pp. 321–331–332.

of view brings us really one step nearer towards the correct interpretation of myth, then it is clear that we need not rely so much on roundabout or symbolic reinterpretations of facts, but can confidently let the facts speak for themselves. It will be obvious to any attentive reader that many of the situations which we understand as direct results of the matrilineal complex could, by artificial and symbolic rehandling, be made to correspond to a patriarchal outlook. The conflict between mother's brother and nephew, who should be natural protectors and always keep common cause, but who often in reality regard each other as one ogre might another, the fight and cannibalistic violence between two brothers, who in tribal law form one body, all this corresponds roughly to analogous conflicts within a patriarchal family. And it is just the difference in the actors, in the cast of the play, which distinguishes the matriarchal from the patriarchal myth. It is the sociological point of view of the tragedy which differs. The foundations of the psycho-analytic explanations of myth we have in no way shaken. We have merely corrected the sociology of this interpretation. That this correction, however, is of extreme importance, and even bears upon fundamental psychological problems, has, I trust, been made sufficiently clear.

Let us pass now to the third class of myth, that which we find at the basis of cultural achievement and magic. Magic plays an extremely important part in everything which these natives do. Whenever they approach any subject which is of vital importance to them and in which they cannot rely solely on their own forces, they summon magic to their aid. To master wind and weather, to ward off dangers in sailing, to secure success in love, ceremonial trading or dancing, the natives perform magic. Black magic and magic of health play a very great rôle in their social life, and in the important economic activities and enterprises, such as gardening, fishing, and the construction of canoes, magic enters as an intrinsic and important element. Now between magic and myth there exists an intimate connexion.

Most of the super-normal power displayed by the heroes in myth is due to their knowledge of magic. Present humanity differs from the great mythical heroes of the past in that nowadays the most effective types of magic have been lost. Could the strong spells and the powerful rites be recovered, men could fly through the air, rejuvenate and thus retain their life for ever, kill people and bring them to life again, be always beautiful, successful, loved and praised.

But it is not only myth which draws its power from magic. Magic is also dependent upon myth. Almost every type of spell and rite has its mythological foundation. The natives tell a story of the past which explains how this magic came into man's possession, and which serves as a warrant of its efficiency. In this lies perhaps the main sociological influence of myth. For myth lives in magic, and since magic shapes and maintains many social institutions, myth exercises its influence upon them.

Let us now pass to a few concrete examples of such myths of magic. It will be best to discuss the question of one detailed case first, and for this I shall choose the myth of the flying canoe already published in *extenso*.[1] This myth is narrated in connexion with the ship-building magic used by the natives. A long story is told about a time when there existed magic which, performed during the construction of a canoe, could make it fly through the air. The hero of this story, the man who was the last—and as it seems also the first—to perform it, is depicted in his rôle of ship-builder and magician. We are told how under his direction a canoe is built; how, on an overseas expedition to the south, it outruns all others, flying through the air while they have to sail; how its owner obtains an overwhelming success in the expedition. This is the happy beginning of the story. Now comes the tragedy. All the men in the community are jealous and full of hatred against the hero. Another incident occurs. He is in

[1] *Op. cit.*, pp. 421 *sqq.*

possession also of a successful garden magic, and of one by which he can also damage his neighbours. In a general drought his garden alone survives. Then all the men of his community determine that he must die. The younger brother of the hero had received from him the canoe magic and the garden magic. So no one thought that by killing the elder brother they would also lose the magic. The criminal deed is performed, and it is done not by any strangers, but by the younger brother of the hero. In one of the versions he and the hero's maternal nephews kill him in a joint attack. In another version again, the story proceeds to tell how, after he has killed his elder brother, he then proceeds to organize the mortuary festivities for him. The point of the story remains in the fact that after the deed was done, and the younger brother tried to apply the magic to a canoe, he found out with dismay that he was not in possession of the full magic, but only of its weaker part. Thus humanity lost the flying magic for ever.

In this myth the matrilineal complex comes powerfully to the fore. The hero, whose duty it is according to tribal law to share the magic with his younger brother and maternal nephew, cheats them, to put it in plain terms, by pretending that he has handed them over all the spells and rites while in reality he only gave up an insignificant fraction. The younger man, on the other hand, whose duty it would be to protect his brother, to avenge his death, to share all his interests, we find at the head of the conspiracy, red-handed with fratricidal murder.

If we compare this mythical situation with the sociological reality we find a strange correspondence. It is the duty of every man to hand over to his maternal nephew or younger brother the hereditary possessions of the family, such as family myth, family magic and family songs; as well as the titles to certain material possessions and economic rites. The handing over of magic has obviously to be done during the life-time of the elder man. The cession of property rights and privileges is also frequently done before his death. It is interesting that such lawful acquisition by a man of the goods which are due to him by inheritance from his maternal uncle or elder brother has always to be done against a type of payment called *pokala*, which frequently is very substantial indeed. It is still more important to note that when a father gives certain properties to his son he always does it for nothing, out of sheer affection. In actual life, the mythological swindle of the younger by the elder brother is also very often paralleled. There is always a feeling of uncertainty, always a mutual suspicion between the two people who in tribal law should be at one in common interests and reciprocal duties as well as in affection. Ever so often when obtaining magic from a man, I became aware that be was himself doubtful whether he had not been cheated out of some of it in receiving it from his uncle or elder brother. Such a doubt was never in the mind of a man who had received his magic as a gift from the father. Surveying the people now in possession of important systems of magic, I find also that more than half of the outstanding younger magicians have obtained their powers by paternal gift and not by maternal inheritance.

Thus in real life, as well as in myth, we see that the situation corresponds to a complex, to a repressed sentiment, and is at cross variance with tribal law and conventional tribal ideals. According to law and morals, two brothers or a maternal uncle and his nephew are friends, allies, and have all feelings and interests in common. In real life to a certain degree and quite openly in myth, they are enemies, cheat each other, murder each other, and suspicion and hostility obtain rather than love and union.

One more feature in the canoe myth deserves our attention: in an epilogue to the myth we are told that the three sisters of the hero are angry with the younger brother because he has killed the elder one without learning the magic. They had already learnt it, however, and, though, being women, they could not build or sail flying canoes, they were able to fly through the air as flying witches. After the crime had been committed they flew away, each of them settling in a different district. In this episode we see the characteristic matrilineal position of woman, who learns magic first before man has acquired it. The sisters also appear as moral guardians of the clan, but their wrath is directed not against the crime, but against the mutilation of clan property. Had the younger brother known the magic before kiling the elder, the three sisters would have lived on happily with him for ever after.

Another fragmentary myth already published deserves our attention,[1] the myth about the origins of salvage magic, in cases of shipwreck. There were two brothers, the elder a man, the younger a dog. One day the senior goes on a fishing expedition, but he refuses to take the younger one with him. The dog, who has acquired the magic of safe swimming from the mother, follows the elder one, diving under water. In the fishing the dog is more successful. In retaliation for the ill-treatment received from the elder brother, the dog changes his clan and bequeaths the magic to his adopted kinsmen. The drama of this myth consists first of all in the favouring by the mother of the second son, a distinctly matrilineal feature, in that the mother here distributes her favours directly, and does not need to cheat the father like her better-known colleague in the Bible, the mother of Esau and Jacob. There is also the typical matrilineal quarrel, the wronging of the younger brother by the elder, and retaliation.

One more important story has to be given here: the legend about the origin of love magic, which forms the most telling piece of evidence with regard to the influence of the matrilineal complex. Among these amorous people the arts of seduction, of pleasing, of impressing the other sex, lead to the display of beauty, of prowess, and of artistic abilities. The fame of a good dancer, of a good singer, of a warrior, has its sexual side, and

[1]   Op. cit. pp. 262–4.

though ambition has a powerful sway for its own sake, some of it is always sacrificed on the altar of love. But above all the other means of seduction the prosaic and crude art of magic is extensively used, and it commands the supreme respect of the natives. The tribal Don Juan will boast about his magic rather than any personal qualities. The less successful swain will sigh for magic: 'If I only knew the real *Kayroiwo*' is the burden of the broken heart. The natives will point to old, ugly, and crippled men who yet have been always successful in love by means of their magic.

This magic is not simple. There is a series of acts, each consisting of a special formula and its rite, which have to be carried out one after the other in order to exercise an increasing charm upon the desired lover. It may be added at once that the magic is carried out by girls to capture an admirer as well as by youths to subdue a sweetheart.

The initial formula is associated with the ritual bathe in the sea. A formula is uttered over the spongy leaves which are used by the natives as a bathing towel to dry and rub the skin. The bather rubs his skin with the bewitched leaves, then throws them into the waves. As the leaves heave up and down so shall the inside of the beloved one be moved by passion. Sometimes this formula is sufficient; if not, the spurned lover will resort to a stronger one. The second formula is chanted over betel-nut, which the lover then chews and spits out in the direction of his beloved. If even this should prove unavailing, a third formula, stronger than the two preceding ones, is recited over some dainty, such as betel-nut or tobacco, and the morsel is given to the desired one to eat, chew, or smoke. An even more drastic measure is to utter the magic into the open palms and attempt to press them against the bosom of the beloved.

The last and most powerful method might, without pushing the simile too far, be described as psycho-analytic. In fact, long before Freud had discovered the predominantly erotic nature of dreams, similar theories were in vogue among the brown-skinned savages of north-west Melanesia. According to their view, certain forms of magic can produce dreams. The wish engendered in such dreams penetrates into waking life and thus the dream-wish becomes realized. This is Freudianism turned upside-down; but which theory is correct and which is errone-ous I shall not try definitely to settle. As regards love magic, there is a method of brewing certain aromatic herbs in coconut oil and uttering a formula over them, which gives them a powerful dream-inducing property. If the magic-maker be successful in making the smell of this brew enter the nostrils of his beloved, she will be sure to dream of him. In this dream she may have visions and undergo experiences which she will inevitably attempt to translate into deeds in actual life.

Among the several forms of love magic that of the *sulumwoya* is by far the most important. A great potency is ascribed to it, and it commands a considerable price if a native wants to purchase the formula and the rite, or if he wants it to be performed on his behalf. This magic is localized in two centres. One of them lies on the eastern shore of the main island. A fine beach of clean coral sand overlooks the open sea towards the west, where beyond the white breakers on the fringing reef there may be seen on a clear day silhouettes of distant raised coral rocks. Among them is the island of Iwa, the second centre of love magic. The spot on the main island, which is the bathing and boating beach of the village of Kumilabwaga, is to the natives almost like a holy shrine of love. There, in the white limestone beyond the fringe of luxuriant vegetation is the grotto where the primeval tragedy was consummated; there on both sides of the grotto are the two springs which still possess the power of inspiring love by ritual.

A beautiful myth of magic and love connects these two spots facing each other across the sea. One of the most interesting aspects of this myth is that it accounts for the existence of love-magic by what to the natives is a horrible and tragic event, an act of incest between brother and sister. In this the story shows some affinity to the legends of Tristan and Isolde, Lancelot and Guinevere, Sigmund and Sigelinde, as well as to a number of similar tales in savage communities.

There lived in the village of Kumilabwaga a woman of the Malasi clan who had a son and a daughter. One day while the mother was cutting out her fibre-petticoat, the son made some magic over herbs. This he did to gain the love of a certain woman. He placed some of the pungent kwayawaga leaves and some of the sweet-scented sulumwoya (mint) into clarified coconut oil and boiled the mixture, reciting the spell over it. Then he poured it into a receptacle made of toughened banana-leaves and placed it in the thatch. He then went to the sea to bathe. His sister in the meantime had made ready to go to the water hole to fill the coconut bottles with water. As she passed under the spot where the magical oil had been put, she brushed against the receptacle with her hair and some of the oil dropped down over her. She brushed it off with her fingers, and then sniffed at them. When she returned with the water she asked her mother, 'Where is the man, where is my brother?' This according to native moral ideas was a dreadful thing to do, for no girl should inquire about her brother, nor should she speak of him as a man. The mother guessed what had happened. She said to herself: 'Alas, my children have lost their minds.'

The sister ran after her brother. She found him on the beach where he was bathing. He was without his pubic leaf. She loosened her fibre skirt and naked she tried to approach him. Horrified by this dreadful sight the man ran away along the beach till he was barred by the precipitous rock which on the north cuts off the Bokaraywata beach. He turned and ran back to the other rock which stands up steep and inaccessible at the southern end. Thus they ran three times along the beach under the shade of the big overhanging trees till the man, exhausted and overcome, allowed his sister to catch hold of him, and the two fell down, embracing in the shallow water of the caressing waves. Then, ashamed and remorseful, but with the fire of their love not quenched, they went to the grotto of Bokaraywata where they remained without food, without drink, and without sleep. There also they died, clasped in one another's arms, and through their linked bodies there grew the sweet-smelling plant of the native mint (sulumwoya).

A man in the island of Iwa dreamt the kirisala, the magical dream of this tragic event. He saw the vision before him. He woke, and said: 'The two are dead in the grotto of Bokaraywata and the sulumwoya is growing out of their bodies. I must go.' He took his canoe; he sailed across the sea between his island and that of Kitava. Then from Kitava he went to the main island, till he alighted on the tragic beach. There he saw the reef-heron hovering over the grotto. He went in and he saw the sulumwoya plant growing out of the lovers' chests. He then went to the village. The mother avowed the shame which had fallen on her family. She gave him the magical formula, which he learned by heart. He took part of the spell over to Iwa and left part of it in Kumilabwaga. At the grotto he plucked off some of the mint, and took it with him. He returned to Iwa, to his island. He said: 'I have brought here the tip of the magic; its roots remain in Kumilabwaga. There it will stay, connected with the bathing passage of Kadiusawasa and with the water of Bokaraywata. In one spring the men must bathe, in the other the women.' The man of Iwa then imposed the taboos of the magic, he prescribed exactly the ritual and he stipulated that a substantial payment should be made to the people of Iwa and Kumilabwaga, when they allowed others to use their magic or to use their sacred spots. There is also a traditional miracle or at least an augury to those who perform the magic on the beach. In the myth this is represented as laid down by the man of Iwa; when the magic is performed and good results can be foreseen two small fish will be seen playing together in the shallow water of the beach.

I have but summarized here this last part of the myth, for its literal form contains sociological claims which are wearisome and degenerate into boastings; the account of the miraculous element usually leads into reminiscences from the immediate past; the ritual details develop into technicalities and the list of taboos into prescriptive homilies. But to the native narrator this last part of practical, pragmatic, and often of personal interest, is perhaps more important than the rest, and the anthropologist has more to learn from it than from the preceding dramatic tale. The sociological claims are contained in the myth, since the magic to which it refers is personal property. It has to be handed over from a fully entitled possessor to one who lawfully acquires it from him. All the force of magic consists in correct tradition. The fact of direct filiation by which the present officiator is linked to the original source is of paramount relevance. In certain magical formulæ the names of all its wielders are enumerated. In all rites and spells the conviction that they are absolutely in conformity with the original pattern is essential. And myth figures as the ultimate source, as the last pattern of this retrogressive series. It is again the charter of magical succession, the starting-point of the pedigree.

In connection with this a few words must be said about the social setting of magic and myth. Some forms of magic are not localized. Here belong sorcery, love magic, beauty magic, and the magic of *Kula*. In these forms filiation is none the less important, although it is not filiation by kinship. Other forms of magic are associated with a given territory, with the local industries of a community, with certain paramount and exclusive claims, vested in a chief and in his capital village. All garden magic belongs here—the magic which must be born of the soil, on which it can only thus be efficacious. Here belongs the magic of the shark and other fishing of a local character. Here also belong certain forms of canoe magic, that of the red shell used for ornaments, and, above all, *waygigi*, the supreme magic of rain and sunshine, the exclusive privilege of the paramount chiefs of Omarakana.

In these types of local magic the esoteric power of words is as much chained to the locality as the group who inhabit the village and wield the magic. The magic thus is not merely local but exclusive and hereditary in a matrilineal kinship group. In these cases the myth of magic must be placed side by side with the myth of local origins as an essentially sociological force welding the group together, supplying its quota to the sentiment of unity, endowing the group with a common cultural value.

The other element conspicuous in the end of the above story and present also in most other magic-myths is the enumeration of portents, auguries, and miracles. It might be said that as the local myth establishes the claims of the group by precedent, so the magical myth vindicates them by miracle. Magic is based upon the belief in a specific power, residing always in man, derived always from tradition.[1] The efficiency of this power is vouched for by the myth, but it has to be confirmed also by the only thing which man ever accepts as final proof, namely practical results. 'By their fruits ye shall know them.' Primitive man is not less eager than the modern man of science to confirm his convictions by empirical fact. The empiricism of faith, whether savage or civilized, consists in miracles. And living belief will always generate miracles. There is no civilized religion without its saints and devils, without its illuminations and tokens, without the spirit of God descending upon the community of the faithful. There is no new-fangled creed, no new religion, whether it be a form of Spiritism, Theosophy, or Christian Science, which cannot prove its legitimacy by the solid fact of supernatural manifestation. The savage has also his thaumatology, and in the Trobriands, where magic dominates all

[1] Op. cit., chapters on 'Magic' and 'Power of Words in Magic', cf. also Ogden and Richards' The Meaning of Meaning, chap. ii.

supernaturalism, it is a thaumatology of magic. Round each form of magic there is a continuous trickle of small miracles, at times swelling into bigger, more conspicuously supernatural proofs, then again, running in a smaller stream, but never absent.

In love magic, for instance, from the continuous boasting about its success, through certain remarkable cases in which very ugly men arouse the passion of famous beauties, it has reached the climax of its miracle-working power in the recent notorious case of incest mentioned above. This crime is often accounted for by an accident similar to that which befell the mythical lovers, the brother and sister of Kumilabwaga. Myth thus forms the background of all present-day miracles; it remains their pattern and standard. I might quote from other stories a similar relation between the original miracle narrated by myth and its repetition in the current miracles of living faith. The readers of *The Argonauts of the Western Pacific* will remember how the mythology of ceremonial trading casts its shadow on modern custom and practice. In the magic of rain and weather, of gardening and of fishing, there is a strong tendency to see the original miracle repeated in an attenuated form in ﹐outstanding miraculous confirmations of magical power.

Finally, the element of prescriptive injunction, the laying down of ritual, taboos, and social regulations crops up towards the end of most mythical narratives. When the myth of a certain magic is told by a wielder of the magic, he naturally will state his own functions as the outcome of the story. He believes himself to be at one with the original founder of the magic. In the love myth, as we have seen, the locality in which the primeval tragedy happened, with its grotto, its beach, and its springs, becomes an important shrine infused with the power of magic. To the local people, who no longer have the exclusive monopoly of magic, certain prerogatives still associated with the spot are of the greatest value. That part of the ritual which still remains bound to the locality naturally occupies their attention. In the magic of rain and sunshine of Omarakana, which is one of the corner-stones of the chief's power, the myth revolves round one or two local features which also figure in present day ritual.

All sexual attraction, all power of seduction, is believed to reside in the magic of love.

In the fishing of shark and of the *kalala*, specific elements of the locality figure also. But even in these stories which do not wed magic to locality, long prescriptions of ritual are either told as an integral part of the narrative or else are put in the mouth of one of the *dramatis personæ*. The prescriptive character of myth shows its essentially pragmatic function, its close association with ritual, with belief, with living culture. Myth has often been described by writers of psycho-analysis as 'the secular dream of the race'. This formula, even as a rough approximation, is incorrect in view of the practical and pragmatic nature of myth just established. It has been necessary barely to touch upon this subject here, for it is treated more fully in another place.[1]

In this work I trace the influence of a matrilineal complex upon one culture only, studied by myself at first-hand in intensive field work. But the results obtained have a much wider application. For myths of incest between brother and sister are of frequent occurrence among matrilineal peoples, especially in the Pacific, and hatred and rivalry between elder and younger brother, or between nephew and maternal uncle, is a characteristic feature of the world's folklore.

[1] 'Myth in Primitive Psychology,' *Psyche Miniatures*, 1926.

Part III

Psycho-analysis and Anthropology

1

THE RIFT BETWEEN PSYCHO-ANALYSIS AND SOCIAL SCIENCE

The psycho-analytic theory of the Œdipus complex was first framed without any reference to the sociological or cultural setting. This was only natural, for psycho-analysis started as a technique of treatment based on clinical observation. It was subsequently expanded into a general account of neuroses; then into a theory of psychological processes in general; finally it became a system by which most phenomena in body and mind, in society and culture were to be explained. Such claims are obviously too ambitious, but even their partial realization could have been possible only through intelligent and whole-hearted co-operation between experts in psycho-analysis and the various other specialists. These latter might have become acquainted with psycho-analytical principles and been led by these into new avenues of research. In turn, they might have placed their special knowledge and their methods at the disposal of psycho-analysts.

Unfortunately the new doctrine was not accorded a benevolent and intelligent reception: on the contrary most specialists either ignored or combated psycho-analysis. As a consequence we find a somewhat rigid and esoteric seclusion on the psycho-analytic side and ignorance of what is without doubt an important contribution to psychology on the other.

This book is an attempt at a collaboration between anthropology and psycho-analysis. Several similar attempts have also been made from the psycho-analytic side, as an example of which I shall take an interesting article by Dr. Ernest Jones.[1] This is of special moment, since it is a criticism of the first part of this book, which appeared as two preliminary articles in 1924.[2] Dr. Jones's essay will serve as a typical illustration of certain differences in the method of approach of anthropologists and of psycho-analysts to the problems of primitive society; it is especially suited for this since the author, in his interpretation of mother-right among the Melanesians, his understanding of the complexity of their legal system and of their kinship organization, reveals his grasp of difficult anthropological questions.

It will be convenient here to give a short summary of the views expressed by Dr. Jones. The purpose of his essay is to give a psycho-analytic explanation of the institution of mother-right and of the ignorance of paternity which obtains among certain primitive peoples. According to the psycho-analyst these two phenomena are not to be taken merely at their face value. Thus savages, when propounding their views on procreation, display symbolism of such an accurate kind 'as to indicate at least an unconscious knowledge of the truth'. And this repressed cognizance of the facts of paternity stands in the closest relationship to the features of the mother-right, since each is actuated by the

---

[1] 'Mother-Right and the Sexual Ignorance of Savages', *International Journal of Psycho-Analysis*, vol. vi, part 2, 1925, pp. 109–30.
[2] 'Psycho-Analysis and Anthropology', *Psyche*, vol. iv.

same motive—the wish to deflect the hatred felt by the growing boy towards his father.

In support of this hypothesis Dr. Jones draws to a considerable extent upon the material from the Trobriand Islands, but differs from my conclusions, notably in regard to the central theme— the determination of the form of the nuclear family complex by the social structure of the particular culture observed. Dr. Jones adheres to Freud's theory of the Œdipus complex as a fundamental—in fact primordial—phenomenon. He is of the opinion that of the two elements which compose it, love for the mother and hatred against the father, the latter is by far the most important in leading to repression. From this an avenue of escape is sought by simply denying the act of birth from the father, 'repudiation of the father's part in coitus and procreation, and consequently softening and deflection of the hatred against him' (p. 122). But the father is not yet disposed of. The 'atti- tudes of awe, dread, respect and suppressed hostility which are inseparable from the idea of the father imago', springing from 'the obsessional ambivalence of savages', have still to be dealt with, so the maternal uncle is chosen, so to speak, as the scape- goat on whom can be heaped all the sins of the older male in authority, while the father can continue a friendly and pleasant existence within the household. Thus we have a 'decomposition of the primal father into a kind and lenient actual father on the one hand and a stern and moral uncle on the other' (p. 125). In other words the combination of mother-right and ignorance protects both father and son from their maternal rivalry and hostility. For Dr. Jones, then, the Œdipus complex is funda- mental; and 'the matrilineal system with its avunculate complex arose, . . . as a mode of defence against the primordial Œdipus tendencies' (p. 128).

All these views will strike the readers of the first two parts of this book as not altogether unfamiliar, and sound in all the essentials.

I am not prepared to endorse unconditionally Dr. Jones's main contention that both mother-right and ignorance of paternity have come into being 'to deflect the hatred towards his father felt by the growing boy' (p. 120). I think this statement requires a fuller testing in the various anthropological provinces. But this view seems to me to be perfectly well in harmony with all the facts which I have discovered in Melanesia, and with any other kinship systems with which I am acquainted through literature. Should Dr. Jones's hypothesis become established by subsequent research, as I think and hope it will be, the value of my own contributions will obviously be very much enhanced. For instead of having drawn attention to a mere accidental constellation, I should have had the good fortune to discover phenomena of universal evolutionary and genetic importance. In a way it seems to me that Dr. Jones's hypothesis is a daring and original extension of my own conclusions, that in mother-right the family complex must be different from the Œdipus complex; that in the matrilineal conditions the hate is removed from the father and placed upon the maternal uncle; that any incestuous temptations are directed towards the sister rather than towards the mother.

Dr. Jones takes, however, not only a more comprehensive point of view, in which I am prepared to follow him; he places, besides, a certain causal or metaphysical stress in that he regards the complex as the *cause*, and the whole sociological structure as the *effect*. In Dr. Jones's essay, as in most psycho-analytic interpretations of folklore, custom and institutions, the universal occurrence of the Œdipus complex is being assumed, as if it existed independently of the type of culture, of the social organization and of the concomitant ideas. Wherever we find in folklore hatred between two males, one of them is interpreted as symbolizing the father, the other the son, irrespective of whether in that society there are any opportunities for a father and son to conflict. Again, all repressed or illicit passion which we find so often in mythological tragedies is due to the incestuous love between mother and son, even though such temptations could be shown to have been eliminated by the type of organization prevalent in that community. Consequently Dr. Jones in the article quoted above maintains that while my results may be correct 'on the purely descriptive plane', the correlation between sociology and psychology on which I insist is 'extremely doubtful' (p. 127). And again that 'if attention is concentrated on the sociological aspects of the data' my view might 'appear a very ingenuous and perhaps even plausible suggestion', but that it was only my 'imperfect attention to the genetic aspects of the problem' which 'has led to a lack of . . . a dimensional perspective, i.e. a sense of value based on intimate knowledge of the unconscious' (p. 128). Dr. Jones arrives at the conclusion, somewhat crushing to me, 'that the opposite of Malinowski's conception is nearer the truth' (ibid).

The radical discrepancy between psycho-analytical doctrine and empirical anthropology or sociology implied in these quotations does not seem to me to exist. I should not like to see psycho-analysis divorced from the empirical science of culture, nor the descriptive work in anthropology deprived of the assistance of psycho-analytical theory. I cannot myself plead guilty of overemphasizing the sociological elements either. I have tried to introduce these factors into the formula of the nuclear complex, but I have in no way minimized the importance of biological, psychological, or unconscious factors.

2

A 'REPRESSED COMPLEX'

My main contention is concisely and adequately summed up by Dr. Jones himself as 'the view that the nuclear family complex varies according to the particular family structure existing in any community. According to him (i.e. to Malinowski) a matrilineal family system arises, for unknown social and economic reasons, and then the repressed nuclear complex consists of brother and sister attraction, with nephew and uncle hatred; when this system is replaced by a patrilineal one, the nuclear complex becomes the familiar Œdipus one' (pp. 127 and 128). All this is a perfectly correct interpretation of my views, though Dr. Jones has gone beyond the scope of my previously published conclusions. As a field-worker I have remained throughout my essay on the 'purely descriptive plane', but in this Part I shall presently take the opportunity of setting forth my genetic views.

As has been already mentioned, the crux of the difficulty lies in the fact that to Dr. Jones and other psycho-analysts the Œdipus complex is something absolute, the primordial source, in his own words the *fons et origo* of everything. To me on the other hand the nuclear family complex is a functional formation dependent upon the structure and upon the culture of a society. It is necessarily determined by the manner in which sexual restrictions are moulded in a community and by the manner in which authority is apportioned. I cannot conceive of the complex as the first cause of everything, as the unique source of culture, of organization and belief; as the metaphysical entity, creative, but not created, prior to all things and not caused by anything else.

Let me quote some more significant passages from Dr. Jones's article in order to indicate the obscurities and contradictions to which I have alluded. They illustrate the type of argument which we meet in the orthodox psycho-analytic discussions of savage custom.

Even where they admittedly cannot be found in actual existence, as in the matrilineal societies of Melanesia, the 'primordial Œdipus tendencies' are still lurking behind: 'The forbidden and unconsciously loved sister is only a substitute for the mother, as the uncle plainly is for the father' (p. 128). In other words the Œdipus complex is merely screened by another one, or painted over by the other complex, in slightly different colours. As a matter of fact, Dr. Jones uses an even stronger terminology and speaks about the 'repression of the complex' and about 'the various complicated devices whereby this repression is brought about and maintained' (p. 120). And here comes the first obscurity. I have always understood that a complex is an actual configuration of attitudes and sentiments partly overt, partly repressed, but actually existing in the unconscious. Such a complex can always be empirically reached by the practical methods of psycho-analysis, by the study of mythology, folklore and other cultural manifestations of the unconscious. If, however, as Dr. Jones seems fully to admit, the attitudes typical of the Œdipus complex cannot be found either in the conscious or unconscious; if, as has been proved, there are no traces of it either in Trobriand folklore or in dreams and visions, or in any other symptoms; if in all these manifestations we find instead the other complex—where is then the repressed Œdipus complex to be found? Is there a sub-unconscious below the actual unconscious and what does the concept of a repressed repression mean? Surely all this goes beyond the ordinary psycho-analytic doctrine and leads us into some unknown fields; I suspect moreover that they are the fields of metaphysics!

Let us turn to the devices by which the repression of the complex is brought about. According to Dr. Jones they consist in a tendency to divorce relationship and social kinship in the various customary denials of actual birth, in the enactment of a ritual birth, in the affectation of ignorance of paternity and so on. I would like to state here at once that in this I am very much in agreement with Dr. Jones's point of view, though I might differ in certain details. Thus I am not quite sure whether I would speak of a 'tendentious denial of physical paternity' since I am firmly convinced that the ignorance of these complicated physiological processes is as natural and direct as is the ignorance of the processes of digestion, secretion, of the gradual bodily decay, in short, of all that happens in the human body. I do not know why we should assume that people on a very low level of culture have received their early revelation about certain aspects of embryology while in all other aspects of natural science they know next to nothing as to the causal connections of phenomena. That, however, the divorce or at least the partial autonomy of biological and social relations under culture is of the greatest importance in primitive society I shall try to demonstrate presently at some length.

In the matter of ignorance of paternity, however, there seems to me a slight discrepancy in Dr. Jones's views. In one place we are told 'there is the closest collateral relationship between ignorance about paternal procreation on the one hand and the institution of mother-right on the other. My view is that both these phenomena are brought about by the same motive; in what chronological relation they stand to each other is another question altogether, which will be considered later. The motive, according to this view, in both cases is *to deflect the hatred towards his father felt by the growing boy*' (p. 120). The point is crucial and yet Dr. Jones does not himself feel quite certain about it. For in another place he tells us that there is no 'reason to suppose that the savage ignorance, or rather repression, of the facts of paternal procreation is a necessary accompaniment of mother-right, though it is evident that it must be a valuable support to the motives discussed above which led to the instituting of mother-right' (p. 130). The relation between the two sentences quoted is not quite clear, and while the latter is not quite correct the former would be more lucid if we were told what the author means by the 'closest collateral relationship'. Does that mean that both ignorance and mother-right are *necessary* effects of the principal cause, i.e., the Œdipus complex, or are they both loosely connected with it? If so what are the conditions under which the necessity to mask the Œdipus complex leads to mother-right and ignorance, and what are the conditions in which it does not lead to these effects? Without such concrete data Dr. Jones's theory is not much more than a vague suggestion.

Having examined the devices, let us have a look at the 'primordial cause'. This, as we know, is the Œdipus complex conceived in an absolute and genetically transcendental manner. Going beyond Dr. Jones's essay to the anthropological contributions of psycho-analysts in general, we learn how the Œdipus complex is supposed to have originally come into being. It originated by the famous totemic crime in the primeval horde.

3

'THE PRIMORDIAL CAUSE OF CULTURE'

Freud's theory of the dramatic beginnings of totemism and taboo, of exogamy and sacrifice, is of great importance in all psycho-analytic writings on anthropology. It cannot be passed over in any essay like the present one, which tries to bring psycho-analytic views into line with anthropological findings. We shall therefore take this opportunity of entering into a detailed critical analysis of the theory.

In his book on *Totem and Taboo* Freud shows how the Œdipus complex can serve to explain totemism and the avoidance of the mother-in-law, ancestor worship and the prohibitions of incest, the identification of man with his totemic animal and the idea of the God Father.[1] In fact the Œdipus complex, as we know, has to be regarded by psycho-analysts as the source of culture, as having occurred before the beginnings of culture, and in his book

[1] S. Freud, *Totem and Taboo*, New York, 1918. The quotations in the text refer to the American Edition.

Freud gives us precisely the hypothesis describing how it actually came into being.

In this Freud takes the cue from two illustrious predecessors, Darwin and Robertson Smith. From Darwin he borrows the idea of 'primal horde' or as it was renamed by Atkinson 'the Cyclopean family'. According to this view the earliest form of family or social life consisted of small groups led and dominated by a mature male who kept in subjection a number of females and children. From another great student, Robertson Smith, Freud received the suggestion about the importance of the totemic sacrament. Robertson Smith considers that the earliest act of religion consisted of a common meal in which the totemic animal was ceremonially eaten by the members of the clan. In later development sacrifice, the almost universal and certainly the most important religious act, emerged from the totemic meal. The taboo forbidding the eating of totemic species at ordinary times constitutes the negative side of the ritual communion. To these two hypotheses Freud added one of his own: the identification of man with the totem is a trait of the mentality common to children, primitives and neurotics, based upon the tendency to identify the father with some unpleasant animal.

In this context we are primarily interested in the sociological side of the theory and I shall quote in full the passage of Darwin's upon which is built Freud's theory. Says Darwin: 'We may indeed conclude from what we know of the jealousy of all male quadrupeds, armed, as many of them are, with special weapons for battling with their rivals, that promiscuous intercourse in a state of nature is extremely improbable . . . If we therefore look back far enough into the stream of time and judging from the social habits of man as he now exists, the most probable view is that he originally lived in small communities, each with a single wife, or if powerful with several, whom he jealously defended against all other men. Or he may not have been a social animal and yet have lived with several wives, like the gorilla; for all the natives agree that only the adult male is seen in a band; when the young male grows up a contest takes place for mastery and the strongest, by killing and driving out the others, establishes himself as the head of the community (Dr. Savage in the *Boston Journal of Natural History*, vol. v, 1845–47). The younger males thus being driven out and wandering about would also, when at last successful in finding a partner, prevent too close inbreeding within the limits of the same family.'[1]

I may at once point out that in this passage Darwin speaks about man and gorillas indiscriminately. Nor is there any reason why we as anthropologists should blame him for this confusion—the least our science can do is to deprive us of any vanities with regard to our anthropoid brethren! But if philosophically the difference between a man and a monkey is insignificant, the distinction between *family* as we find it among the anthropoid apes and the organized human family is of extreme importance for the sociologist. He has to discriminate clearly between animal life in the state of nature and human life under culture. To Darwin, who was developing a biological argument against the hypothesis of promiscuity, the distinction was irrelevant. Had he been dealing with the origins of culture, had he tried to define the moment of its birth, the line of distinction between nature and culture would have been all-important. Freud who, as we shall see, actually does attempt to grasp and to render the 'great event with which culture began', fails completely in his task in that he loses sight of this line of division and places culture in conditions in which, *ex hypothesi*, it cannot exist. Darwin speaks, moreover, only about the *wives* of the leader of the herd, and not of any other females. He also states that the excommunicated young males succeed finally in finding a partner and do not trouble any more about their parental family. On

---

[1] S. Freud, *Totem and Taboo*, 1918, pp. 207–8, quoted from Darwin, 'The Descent of Man,' vol. ii, chapter 20, pp. 603–4.

both these points Freud substantially modifies the Darwinian hypothesis.

Let me quote the words of the master of psycho-analysis in full so as to substantiate my criticism. Says Freud: 'The Darwinian conception of the primal horde does not, of course, allow for the beginnings of totemism. There is only a violent, jealous father who keeps all the females for himself and drives away the growing sons' (p. 233). As we see, the old male is made to keep *all the females* for himself while the expelled sons remain somewhere in the neighbourhood, banded together, in order to be ready for the hypothetical event. And indeed a crime is conjured up before our eyes as bloodcurdling as it is hypothetical, yet of the greatest importance in the history of psycho-analysis, if not of humanity! For according to Freud it is destined to give birth to all future civilization. It is 'the great event with which culture began and which ever since has not let mankind come to rest'; it is the 'deed that was in the beginning'; it is the 'memorable, criminal act with which . . . began social organization, moral restrictions and religion' (pp. 234, 239, 265). Let us hear the story of this primordial cause of all culture.

'One day the expelled brothers joined forces, slew and ate the father, and thus put an end to the father horde. Together they dared and accomplished what would have remained impossible for them singly. Perhaps some advance in culture, like the use of a new weapon, had given them the feeling of superiority. Of course these cannibalistic savages ate their victim. This violent primal father had surely been the envied and feared model for each of the brothers. Now they accomplished their identification with him by devouring him and each acquired a part of his strength. The totem feast, which is perhaps mankind's first celebration, would be the repetition and commemoration of this memorable . . . act . . .' (p. 234).

This is the original act of human culture and yet in the middle of the description the author speaks about 'some advance in culture', about 'the use of a new weapon', and thus equips his pre-cultural animals with a substantial store of cultural goods and implements. No material culture is imaginable without the concomitant existence of organization, morality, and religion. As I shall show presently, this is not a mere quibble but it goes to the very heart of the matter. We shall see that the theory of Freud and Jones tries to explain the origins of culture by a process which implies the previous existence of culture and hence involves a circular argument. A criticism of this position will in fact naturally lead us right into the analysis of cultural process and of its foundations in biology.

4

THE CONSEQUENCES OF THE PARRICIDE

Before we pass a detailed criticism on this theory, however, let us patiently hear all that Freud has to tell us in this matter—it is always worth while listening to him. '. . . the group of brothers banded together were dominated by the same contradictory feelings towards the father which we can demonstrate as the content of ambivalence of the father complex in all our children and in neurotics. They hated their father who stood so powerfully in the way of their sexual demands and their desire for power, but they also loved and admired him. After they had satisfied their hate by his removal and had carried out their wish for identification with him, the suppressed tender impulses had to assert themselves. This took place in the form of remorse, a sense of guilt was formed which coincided here with the remorse generally felt. The dead now became stronger than the living had been, even as we observe it today in the destinies of men. What the father's presence had formerly prevented they themselves now prohibited in the psychic situation of "subsequent obedience" which we know so well from psycho-analysis. They undid their deed by declaring that the killing of the father substitute, the totem, was not allowed, and renounced the fruits of their deed by denying themselves the liberated women. Thus they created the two fundamental taboos of totemism out of the sense of guilt of the son, and for this very reason these had to correspond with the two repressed wishes of the Œdipus complex. Whoever disobeyed became guilty of the two only crimes which troubled primitive society' (pp. 235–6).

We see thus the parricidal sons immediately after the act of murder engaged in laying down laws and religious taboos, instituting forms of social organization, in brief moulding cultural forms which will be handed on far down the history of mankind. And here again we are faced by the dilemma: did the raw material of culture exist already—in which case the 'great event' could not have created culture as it is supposed by Freud to have done, or was culture at the time of the deed not yet in existence—in which case the sons could not have instituted sacraments, established laws and handed on customs.

Freud does not completely ignore this point, though he hardly seems to have recognized its crucial importance. He anticipates the question as to the possibilities of a lasting influence of the primeval crime and of its enduring action across successive generations of man. To meet any possible objections Freud summons to his aid another hypothesis: '. . . it can hardly have escaped anyone that we base everything upon the assumption of a psyche of the mass in which psychic processes occur as in the psychic life of the individual' (p. 259). But this assumption of a collective soul is not sufficient. We have to endow this comprehensive entity also with an almost unlimited memory. '. . . we let the sense of guilt for a deed survive for thousands of years, remaining effective in generations which could not have known anything of this deed. We allow an emotional process such as might have arisen among generations of sons that had been ill-treated by their fathers, to continue to new generations which had escaped such treatment by the very removal of the father' (p. 259).

Freud is somewhat uneasy about the validity of this assumption but an *argumentum ad hominem* is ready at hand. Freud assures us that however daring his hypothesis '. . . we ourselves do not have to carry the whole responsibility for such daring' (p. 260). Not only that: the writer lays down a universal rule for anthropologists and sociologists. 'Without the assumption of a mass psyche, or a continuity in the emotional life of mankind which permits us to disregard the interruptions of psychic acts through the transgression of individuals, social psychology could not exist at all. If psychic processes of one generation did not continue in the next, if each had to acquire its attitude towards life afresh, there would be no progress in this field and almost no development' (p. 260). And here we touch on a very important point: the methodological necessity of the figment of a collective soul. As a point of fact no competent anthropologist now makes any such assumption of 'mass psyche', of the inheritance of acquired 'psychic dispositions', or of any 'psychic continuity' transcending the limits of the individual soul.[1] On the other hand anthropologists can clearly indicate what the medium is in which the experiences of each generation are deposited and

[1] All the anthropological authorities, for instance, upon whom Freud bases his work, Lang, Crawley, Marett, never once in their analysis of custom, belief, and institution have employed such or a similar concept. Frazer above all rules this conception consciously and methodically out of his work (personal communication). Durkheim, who verges upon this metaphysical fallacy, has been criticized on this point by most modern anthropologists. Leading sociologists such as Hobhouse, Westermarck, Dewey, and social anthropologists such as Lowie, Kroeber, Boas, have consistently avoided the introduction of 'the collective sensorium'. For a searching and destructive criticism of certain attempts at a sociological use of 'mass psyche' compare M. Ginsberg, 'The Psychology of Society' (1921).

stored up for successive generations. This medium is that body of material objects, traditions, and stereotyped mental processes which we call culture. It is super-individual but not psychological. It is moulded by man and moulds him in turn. It is the only medium in which man can express any creative impulse and thus add his share to the common stock of human values. It is the only reservoir from which the individual can draw when he wants to utilize the experiences of others for his personal benefit. A fuller analysis of culture to which we shall presently pass will reveal to us the mechanism by which it is created, maintained, and transmitted. This analysis will also show us that the *complex* is the natural by-product of the coming into existence of culture.

It will be obvious to any reader of Dr. Jones's article that he fully adopts Freud's hypothesis about the origins of human civilization. From the passages previously quoted it is clear that to him the Œdipus complex is the origin of everything. Hence it must be a pre-cultural formation. Dr. Jones even more explicitly commits himself to Freud's theory in the following passages: 'Far from being led by consideration of the subject, as Malinowski was, to abandon or revise Freud's conception of the "primal horde" (Atkinson's "cyclopean family"), it seems to me, on the contrary, that this conception furnishes the most satisfactory explanation of the complicated problems which we have been discussing' (p. 130). Dr. Jones also is in full agreement with the racial memory of the original crime, for he speaks about the 'inheritance of impulses dating from the primal horde' (p. 121).

5

THE ORIGINAL PARRICIDE ANALYSED

Let us examine now point by point the hypotheses of Freud and Jones. The hypothesis of the 'primeval horde' has in itself nothing objectionable to the anthropologist. We know that the earliest form of human and pre-human kinship was the family based on marriage with one or more females. In accepting the Darwinian view of kinship, psycho-analysis has discarded the hypotheses of primitive promiscuity, group marriage and sexual communism, and in this it has the full support of competent anthropologists. But as we have seen, Darwin made no explicit distinction between the animal and the human status, and Freud in his reconstruction of Darwin's argument obliterated whatever distinction was implied in the great naturalist's account. We have to enquire therefore into the constitution of the family at the anthropoid end of the human level of development. We have to ask the question: What are the bonds of union within the family, before it became human and after? What is the difference between animal and human kinship; between the anthropoid family in the state of nature and the earliest type of human family under conditions of culture?

The pre-human anthropoid family was united by instinctive or innate bonds, modified by individual experience but not influenced by tradition, for animals have no language, no laws, no institutions. In the state of nature the male and female mate, driven by the selective sexual impulse operating at the time of rut and at that time only. After the impregnation of the female, a new impulse leads to the establishment of common life, the male acting as protector and guarding over the process of pregnancy. With the act of birth, the maternal impulses of suckling, tending, and caring for the offspring appear in the female, while the male responds to the new situation by providing food, keeping watch, and if need be engaging in dangerous combats in the defence of the family. Considering the protracted growth and slow ripening of the individual among the anthropoid apes, it is indispensable for the species that parental love should arise in both male and female and last for some time after birth until the new individual is ready to look after himself. As soon as he is mature there is no biological need to keep the family together. As we shall see, this need arises in culture, where for the sake of co-operation the members of the family must remain united; while for the sake of handing on tradition the new generation must remain in contact with the previous one. But in the pre-human Cyclopean family as soon as the male or female children became independent they would naturally leave the horde.

This is what we find empirically in every simian species. This subserves the interests of the species and has therefore to be assumed on general principles. It also tallies with all which we can infer from our general knowledge of animal instincts. We find also in most higher mammals that the old male leaves the herd as soon as he is past full vigour and thus makes room for a younger guardian. This is serviceable for a species, for as with man temper in animals does not improve with age, and an old leader is less useful and more liable to create conflict. In all this we see that the working of instincts in the condition of nature leaves no room for special complications, inner conflicts, suspended emotions or tragical events.

Family life in the highest animal species is thus cemented and governed by innate emotional attitudes. Where the biological need arises there also appear the appropriate mental responses. When the need ceases the emotional attitude disappears. If we define instinct as a pattern of behaviour in direct response to a situation, a response accompanied by pleasurable feelings—then we can say that animal family life is determined by a chain of linked instincts: courtship, mating, common life, tenderness towards infants and mutual help of the parents. Each of these links follows the other, releasing it completely, for it is character-istic of such concatenations of instinctive responses that each new situation requires a new type of behaviour and a new emo-tional attitude. Psychologically it is very important to realize that each new response replaces and obliterates the old emotional attitude; that no traces of the previous emotion are carried over into the new one. While governed by a new instinct the animal is no more in the throes of a previous one. Remorse, mental con-flict, ambivalent emotion—these are cultural, that is human, and not animal responses. The working of instincts, the unrolling of instinctive sequences, may be more or less successful, accom-panied by more or less friction, but it does not leave any room for 'endopsychic tragedies'.

What is the importance of all this in respect to the hypothesis of primeval crime? I have pointed out repeatedly that the Great Tragedy has been placed by Freud at the threshold of culture and as its inaugural act. Putting aside the several direct quotations from Freud and Jones—and these could be easily multiplied—it is important to realize that this is an assumption indispensable to their theories: all their hypotheses would collapse if we do not make culture begin with the Totemic Parricide. To the psycho-analyst the Œdipus complex is, as we know, the foundation of all culture. This must mean to them not only that the complex governs all cultural phenomena but also that it preceded them all temporarily. The complex is the *fons et origo* out of which there has grown the totemic order, the first elements of law, the beginnings of ritual, the institution of mother-right, everything in fact which to the general anthropologist and to the psycho-analyst counts as the first elements of culture. Dr. Jones objects, moreover, to my attempt at tracing any cultural causes of the Œdipus complex just because this complex antedates all culture. But it is obvious that if the complex has preceded all cultural phenomena, then *a fortiori* the totemic crime, which is the cause of the complex, must be placed still further back.

After having thus established that the event must have happened before culture, we are faced with the other alternative of our dilemma: could that totemic crime have happened in the state of nature? Could it have left traces in tradition and culture, which *ex hypothesi* did not exist at that time? As indicated above, we would have to assume that by one act of collective parricide the Ape had attained culture and become Man. Or again, that by the same act they acquired the so-called racial memory, a new super-animal endowment.

Let us analyze this now in more detail. In the family life of a pre-human anthropoid species each link in the chain of instincts is released as soon as it ceases to be serviceable. Past instinctive attitudes leave no active traces, and neither conflict nor complex attitudes are possible. These assertions should, I submit, be further tested by the student of animal psychology, but they embody all that we know about the subject. If this be so, however, we have to challenge the premises of Freud's Cyclopean hypotheses. Why should the father have to expel the sons if they naturally and instinctively are inclined to leave the family as soon as they have no more need of parental protection? Why should they lack females, if from other groups, as well as from their own, adult children of the other sex have also to come out? Why should the young males remain hanging around the parental horde, why should they hate the father and desire his death? As we know they are glad to be free and they have no wish to return to the parental horde. Why should they finally even attempt or accomplish the cumbersome and unpleasant act of killing the old male, while by merely waiting for his retirement they might gain a free access to the horde should they so desire?

Each of these questions challenges one of the unwarranted assumptions implied in Freud's hypothesis. Freud in fact burdens his Cyclopean family with a number of tendencies, habits, and mental attitudes which would constitute a lethal endowment for any animal species. It is clear that such a view is untenable on biological grounds. We cannot assume the existence in the state of nature of an anthropoid species in which the most important business of propagation is regulated by a system of instincts hostile to every interest of the species. It is easy to perceive that the primeval horde has been equipped with all the bias, maladjustments and ill-tempers of a middle-class European family, and then let loose in a prehistoric jungle to run riot in a most attractive but fantastic hypothesis.

Let us yield, however, to the temptation of Freud's inspiring speculations and admit for the sake of the argument that the primeval crime had been committed. Even then we are faced by insurmountable difficulties in accepting the consequences. As we saw, we are asked to believe that the totemic crime produces remorse which is expressed in the sacrament of endocannibalistic totemic feast, and in the institution of sexual taboo. This implies that the parricidal sons had a conscience. But conscience is a most unnatural mental trait imposed upon man by culture. It also implies that they had the possibilities of legislating, of establishing moral values, religious ceremonies and social bonds. All of which again it is impossible to assume or imagine, for the simple reason that *ex hypothesi* the events are happening in pre-cultural milieu, and culture, we must remember, cannot be created in one moment and by one act.

The actual transition from the state of nature into that of culture was not done by one leap, it was not a rapid process, certainly not a sharp transition. We have to imagine the early developments of the first elements of culture—speech, tradition, material inventions, conceptual thought—as a very laborious and very slow process achieved in a cumulative manner by infinitely many, infinitely small steps integrated over enormous stretches of time. This process we cannot try to reconstruct in detail, but we can state the relevant factors of the change, we can analyze the situation of early human culture and indicate within certain limits the mechanism by which it came about.

To sum up our critical analysis: we have found that the totemic crime must have been placed at the very origins of culture; that it must be made the first cause of culture if it is to have any sense at all. This means that we have to assume the crime and its consequences as happening still in the state of nature, but such an assumption involves us in a number of contradictions. We find that there is in reality a complete absence of motive for a parricidal crime, since the working of instincts is in animal conditions well adjusted to the situation; since it leads to conflicts but not to repressed mental states; since concretely the sons have no reason for hating their father after they have left the horde. In the second place we have seen that in the state of nature there is also a complete absence of any means by which the consequences of the totemic crime could have been fixed into the cultural institutions. There is a complete absence of any cultural medium in which ritual, laws, and morals could have been embodied.

Both objections could be summarized in the verdict that it is impossible to assume origins of culture as one creative act by which culture, fully armed, springs into being out of one crime, cataclysm or rebellion.

In our criticism we have concentrated our attention on what appears to be the most fundamental objection to Freud's hypothesis, an objection connected with the very nature of culture and of cultural process. Several other objections of detail could be registered against this hypothesis but they have been already set forth in an excellent article of Professor Kroeber's in which the anthropological as well as the psycho-analytical inconsistencies of the hypothesis are lucidly and convincingly listed.[1]

There is, however, one more capital difficulty in which psycho-analysis involves itself by its speculations on totemic origins. If the real cause of the Œdipus complex and of culture into the bargain is to be sought in that traumatic act of birth by parricide; if the complex merely survived in the 'race memory of mankind'—then the complex ought obviously to wear out with time. On Freud's theory the Œdipus complex should have been a dreadful reality at first, a haunting memory later, but in the highest culture it should tend to disappear.

This corollary seems inescapable, but there is no need of driving it home dialectically, for Dr. Jones gives it a full and lucid expression in his article. According to him patriarchy, the social organization of the highest cultures, marks indeed the happy solution of all the difficulties due to the primeval crime.

'The patriarchal system, as we know it, betokens acknowledging the supremacy of the father and yet the ability to accept this even with affection, without having recourse to a system either of mother-right or of complicated taboos. It means the taming of man, the gradual assimilation of the Œdipus complex. At last man could face his real father and live with him. Well

---

[1] 'Totem and Taboo, an Ethnologic Psychoanalysis,' *American Anthropologist*, 1920, pp. 48 *seqq.*

might Freud say that the recognition of the father's place in the family signified the most important progress in cultural development.'[1]

Thus Dr. Jones, and on his authority Freud himself, has drawn the inevitable consequence. They admit that within their scheme patriarchal culture—the one most distant from the original course of the complex—is also the one where the gradual assimilation of the 'Œdipus complex' has been achieved. This fits perfectly well into the scheme of *Totem and Taboo*. But how does it fit into the general scheme of psycho-analysis and how does it bear the light of anthropology?

As to the first question, was not the existence of the Œdipus complex discovered in one of our modern patriarchal societies? Is this complex not day by day being re-discovered in the count-less individual psycho-analyses carried on all over the modern patriarchal world? A psycho-analyst should no doubt be the last to answer these questions in the negative. The Œdipus complex does not seem to have been so well 'assimilated' after all. Even if it be admitted that a great deal of exaggeration exists in psycho-analytic findings we have ordinary sociological observation to vindicate the claim of psycho-analysis on this point. But the psycho-analyst cannot have it both ways. He cannot try to cure most ills of the individual mind and of society by dragging their family maladjustments out of the sub-conscious, while at the same time he cheerfully assures us that 'the supremacy of the father is fully acknowledged in our society' and that it is accepted 'even with affection'. Indeed, extreme patriarchal institutions in which *patria potestas* is carried to its bitter end are the very soil for typical family maladjustments. The psycho-analysts have been busy proving that to us from Shakespeare and the Bible, from Roman history and from Greek mythology. Did not the very eponymous hero of the complex—if such an

---

[1] Jones, *loc cit.*, p. 130.

extension of the term be allowed—live in a society pronouncedly patriarchal? And was not his tragedy based on the
father's jealousy and superstitious fear—motives which, by the
way, are typically sociological? Could the myth or the tragedy
unfold before us with the same powerful and fatal effect, unless
we felt the puppets moved by a patriarchal destiny?

Most modern neuroses, the dreams of patients, the myths of
Indo-Germanic peoples, our literatures and our patriarchal
creeds have been interpreted in terms of the Œpidus complex—
i.e. under the assumption that in pronounced father-right the
son never recognizes 'the father's place in the family'; that he
does not like to 'face his real father'; that he is unable to 'live
with him' in peace. Surely psycho-analysis as theory and as
practice stands and falls with the truth of the contention that our
modern culture suffers from the maladjustments covered by the
term Œdipus complex.

What has anthropology to say about the optimistic view
expressed in the passage quoted above? If the patriarchal régime
means the happy solution of the Œdipus complex, the stage
when man could face his father, and so on—then where on earth
does the complex exist in an unassimilated form? That it is
'deflected' under mother-right has been proved in the first two
parts of this book and it has been independently re-vindicated by
Dr. Jones himself. Whether the Œdipus complex in its full splen-
dour does exist in a culture never studied empirically from this
point of view is an idle question. The object of the present work
has been partly to stimulate field workers to further research.
What such an empirical study might or might not reveal I for
one shall not try to foretell. But it seems to me that to deny the
problem, to cover it up with an obviously inadequate assumption, and to obliterate as much as has been already done towards
its solution, is not to render a service either to anthropology or
to psycho-analysis.

I have pointed out a series of contradictions and obscurities in the psycho-analytic approach to this question, taking Dr. Jones's interesting contribution as my main text. Such inconsistencies are: the idea of a 'repressed complex'; the assertion that mother-right and ignorance of paternity are correlated and yet independent; the view that patriarchy is a happy solution of the Œdipus complex as well as its cause. All these discrepancies centre in my opinion round the doctrine that the Œdipus complex is the *vera causa* of social and cultural phenomena instead of being the result; that it originated in the primeval crime; that it continued in racial memory as a system of inherited, collective tendencies.

I would like to indicate just one more point. Taken as a real historical fact, that is one which has to be placed in space and time and concrete circumstance, how is the primitive parricide to be imagined? Have we to assume that once upon a time, in one super-horde, at one spot, one crime had been committed? That this crime then created culture and that this culture spread all over the world by primeval diffusion, changing apes into men wherever it reached? This assumption falls to the ground as soon as formulated. The alternative is equally difficult to imagine: it is a sort of epidemic of minor parricides occurring all over the world, each horde going on with its Cyclopean tyranny and then breaking into crime and thus into culture. The more we look at the hypothesis concretely, the more we try to elaborate it, the less do we feel inclined to treat it as anything but a 'Just-so story', as Professor Kroeber has called it, an appellation not resented by Freud himself.[1]

---

[1] Compare Freud's *Group Psychology and the Analysis of the Ego*, S. Freud, 1922, p. 90. The name of Professor Kroeber is misspelled into "Kroeger" right through all the successive editions. One might inquire what is the psycho-analytic cause of this lapse on the principle developed in *The Psychopathology of Everyday Life*, that no mistake is without its motive. It is almost unpardonable that this misprint of the name of a leading American scholar has been carried over into the American translation of Freud's book!

6

COMPLEX OR SENTIMENT?

I have, up to now, used the word 'complex' to denote the typical attitudes towards members of the family. I have even retrimmed it into a new expression, the nuclear family complex, which is intended to be a generalization, applicable to various cultures, of the term Œdipus complex, whose applicability, I maintain, is restricted to the Aryan, patriarchal society. But, in the interests of scientific nomenclature, I shall have to sacrifice this new compound, nuclear family complex, for not only is it advisable never to introduce new terms, but it is always a laudable act to expurgate science of any terminological intruder, if it can be proved that it is jumping the claim of one already established. I believe that the word 'complex' carries with it certain implications which make it altogether unsuitable, except as a scientific colloquialism—what the Germans call Schlagwort. At the least we must make clear what we mean by it.

The word 'complex' dates from a certain phase of psychoanalysis when this was still in close association with therapy, when it was in fact not much more than a method of treating neurosis. 'Complex' meant the pathogenous, repressed emotional attitude of the patient. But it has now become questionable whether in general psychology one can sever and isolate the repressed part of a man's attitude towards a person, and treat it separately from the non-repressed elements. In our study, we have found that the various emotions which constitute the attitude towards a person are so closely connected and intertwined that they form a closely knit organic insoluble system. Thus, in relation to the father, the feelings which make up veneration and idealization are essentially bound up with the dislike, hatred or scorn which are their reflections. These negative feelings are in fact partly reactions to an over-strained exaltation of the father, shadows cast into the unconscious by the too glaring idealization of the non-ideal father. To sever the shadow from the part which is in the 'foreconscious' and that which is in the unconscious is impossible; they are indissolubly connected. The psycho-analyst in his consultation room can perhaps neglect the open, obvious elements of the attitude which contribute nothing further to the malady; he can isolate the repressed ones and make of them an entity, calling it a complex. But as soon as he leaves his neurotic patients and enters the lecture room with a general psychological theory, he might as well realize that complexes do not exist, that certainly they do not lead an independent existence in the unconscious and that they are only part of an organic whole, of which the essential constituents are not repressed at all.

As a sociologist I am not here concerned with the pathological results, but with their normal, ordinary foundations. And, although it was better to leave this theoretical analysis till now, when we can substantiate it by fact, yet throughout our account of family influences, I have clearly indicated the 'foreconscious' as well as the unconscious elements. Psycho-analysis has the great merit of having shown that the typical sentiments towards father and mother include negative as well as positive elements; that they have a repressed portion, as well as one above the surface of consciousness. But this must not make us forget that both portions are equally important.

Since we see that the conception of an isolated repressed attitude is useless in sociology, we must try to gain a clear vision of how we can generalize it and with what psychological doctrines we should link up our conception of what we had hitherto called 'nuclear family complex', and which includes besides 'unconscious' also overt elements. I have indicated clearly that certain new tendencies of modern psychology have a special affinity to psycho-analysis. I meant, of course, the very important advance of knowledge about our emotional life, inaugurated by Mr. A. F. Shand in his theory of sentiments and developed later by Stout, Westermarck, McDougall, and a few others. Mr. Shand was the first to realize that emotions cannot be treated as loose elements, unconnected and unorganized, floating in our mental medium to make now and then an isolated and accidental appearance. His theory, as well as all the newer work on emotions, is based on the principle first enunciated by himself: namely, that our emotional life is definitely co-ordinated with the environment; that a number of things and persons claim our emotional responses. Round each person or object the emotions are organized into a definite system—the love or hate or devotion we feel for a parent, a country or a life-pursuit. Such a system of organized emotions Mr. Shand calls a sentiment. The ties which bind us to the various members of our family, patriotism, ideals of truth, righteousness, devotion to science—all these are sentiments. And the life of every man is dominated by a limited number of these sentiments. The theory of sentiments was first outlined by Mr. Shand in one or two short essays which must be regarded as epoch-making, and which later were expanded into a large volume.[1] In his book, Mr. Shand assumes

---

[1] 'Character and the Emotions', in *Mind*, new series, vol. i. and *The Foundations of Character*, 1st edn, 1917.

an innate predisposition for a few systems such as love and hate, into each of which there enter a number of emotions. Every emotion again is to Mr. Shand a complex type of mental response to a definite type of situation, so that every emotion has at its command a number of instinctive reactions. Mr. Shand's theory of sentiments will always remain of paramount importance for the sociologist, since social bonds as well as cultural values are sentiments standardized under the influence of tradition and culture. In the study of family life, as it develops in to different civilizations, we have given a concrete application of the Shandian principles, the theory of sentiments with reference to a definite social problem. We have seen how the child's attitude towards the most important items in his environment is gradually formed, and we have examined the influences which contribute towards its formation. The correction and addition which psycho-analysis has allowed us to make to Mr. Shand's theory is the consideration of the repressed elements of a sentiment. But these repressed elements cannot be isolated into a water-tight compartment, and they cannot as a 'complex' be regarded as something different and distinguishable from the 'sentiment'. We see, therefore, that the theory to which we must attach our results in order to put them on a sound theoretical basis is Shand's theory of the sentiments, and that instead of speaking of a 'nuclear complex' we should have to speak of the family sentiments, of kinship ties, typical of a given society.

The attitudes or sentiments towards father, mother, sister and brother do not grow up isolated, detached from one another. The organic, indissoluble unity of the family welds also the psychological sentiments towards its members into one connected system. This is shown very clearly by our results. Thus the expression 'nuclear family complex' is equivalent to the conception of a correlated system of sentiments, or, shortly, of a configuration of sentiments, typical in a patriarchal or a matriarchal society.

Part IV

Instinct and Culture

1

THE TRANSITION FROM NATURE TO CULTURE

In the foregoing part of this book, in which we have been mainly concerned with the discussion of certain psycho-analytic views, our results have been mainly critical: we have tried to establish the principle that in a pre-cultural condition there is no medium in which social institutions, morals, and religion could be moulded; that there is no memory mechanism by which to maintain and to transmit the institutions after they have been established. The position reached is perhaps unassailable to those who really understand the crucial fact that culture cannot be created by one act or in one moment and that institutions, morals, and religion could not be conjured up, even by the greatest cataclysm, among animals who have not yet emerged from the state of nature. But naturally we are not satisfied merely with denying but also want to affirm. We do not merely wish to point out mistakes, but we want to throw light on the actual process. To this end we have to analyze the relation between cultural and natural processes.

The type of behaviour under culture differs essentially from animal behaviour in the state of nature. Man, however simple his culture, disposes of a material outfit of implements, weapons, domestic chattels; he moves within a social milieu which gives him help and controls him in turn; he communicates by speech and thus develops concepts of a rational religious and magical character. Thus man disposes of a body of material possessions, lives within a type of social organization, communicates by language, and is moved by systems of spiritual values. These are perhaps the four main headings under which we usually classify the body of man's cultural achievements. Thus culture appears to us when we meet it as a fact already accomplished. And let us clearly and explicitly recognize that we never can observe it *in statu nascendi*. Nor is it at all profitable to manufacture hypotheses about the 'original events of cultural birth'. What can we do then in trying to reflect upon the beginnings of human culture, that is if we want to do it without having recourse to any extravagant hypotheses or unwarrantable assumptions? There is one important thing to do, namely, to indicate what part various factors of cultural development have played in the process; what they imply in psychological modification of man's endowment, and in what way non-psychological elements can influence this endowment. For the factors of cultural development are intertwined and essentially dependent upon each other, and while we have no knowledge and no indications about sequences in development, while in all speculations about beginnings the element of time entirely escapes our intellectual control, we can yet study the correlations of the factors and thus gain a great deal of information. We have to study these correlations in full cultural development, but we can trace them back into more and more primitive forms. If we thus arrive at a fixed scheme of dependence, if certain lines of correlation appear in all cultural phenomena, we can say that any hypothesis which violates these laws must be considered void. More than this: if the laws of all cultural process disclose to us the paramount influence of certain factors, we must assume that these factors have also been controlling the origins of culture. In this sense the concept of origins does not imply priority in time or causal effectiveness, but merely indicates the universal presence of certain active factors at all stages of development, hence also at the beginning.

Let us start with the recognition that the main categories of culture must from the very outset have been intertwined and simultaneously at work. They could not have originated one after the other, and they cannot be placed in any scheme of temporal sequence. Material culture, for instance, could not have come into being before man was able to use his implements in traditional technique which, as we know, implies the existence of knowledge. Knowledge again and tradition are impossible without conceptual thought and language. Language, thought, and material culture are thus correlated, and must have been so at any stage of development, hence also at the beginnings of culture. The material arrangements of living, again, such as housing, household implements, means of carrying on daily life, are essential correlates and prerequisites of social organization. The hearth and the threshold not only symbolically stand for family life, but are real social factors in the formation of kinship bonds. Morals, again, constitute a force without which man could not battle against his impulses or even go beyond his instinctive endowment, and that he has to do constantly under culture even in his simplest technical activities. It is the changes in instinctive endowment which interest us most in this context, for here we touch the question of repressed drives, of modified impulsive tendencies, that is, the domain of the 'unconscious'. I shall try to show that the neglect to study what happens to human instincts under culture is responsible for the fantastic hypothesis advanced to account for the Œdipus complex. It will be my aim to show that the beginning of culture implies the repression of instincts, and that all the essentials of the Œdipus complex or any other 'complex' are necessary by-products in the process of the gradual formation of culture.

To this end I shall try to show that between the human parent and child under conditions of culture there must arise incestuous temptations which are not likely to occur in animal families governed by true instincts. I shall also establish that these temptations have to be met and ruthlessly repressed in mankind, since incest and organized family life are incompatible. Again, culture implies an education which cannot be carried on without coercive authority. This authority in human society is supplied within the family by the father, and the attitude between father and son gives rise to suppressed hatred and other elements of the complex.

2

THE FAMILY AS THE CRADLE
OF NASCENT CULTURE

The fundamental change in the mechanism of instinctive responses has to be studied upon the very subject matter of our present inquiry: the early forms of family life and the transition between animal and human family. Upon the human family are focussed all psycho-analytic interests and the family is, in the opinion of an anthropological school to which the writer belongs, the most important group in primitive societies.[1] The

[1] It is clear that in this statement, as throughout the book, I imply that the typical form of the human family is based on monogamous marriage. The wide prevalence of monogamy in all human societies is also assumed by Dr. Lowie in his *Primitive Society* (see especially chapter iii). A very interesting and important contribution to the problem is to be found in Capt. Pitt-Rivers's *Contact of Races and Clash of Culture*, 1927 (see especially chapters viii, secs. 1, 2, 3, and xi, sec. 1). Capt. Pitt-Rivers urges the biological and sociological importance of polygyny at the lower levels of culture. Without fully accepting his view, I admit that the problem will have to be rediscussed from the point of view advanced by him. I still maintain, however, that the importance of polygyny is to be found in the rôle which it plays in differentiating the higher

following comparison of courtship, mating, matrimonial relations and parental cares in animal and human societies respectively, will show in what sense the family must be considered as the cell of society, as the starting point of all human organization.

There is one point which must be settled before we can conveniently proceed with our argument. Very often it is assumed by anthropologists that humanity developed from a gregarious simian species and that man inherited from his animal ancestors the so-called 'herd instincts'. Now this hypothesis is entirely incompatible with the view here taken that common sociability develops by extension of the family bonds and from no other sources. Until it has been shown that the assumption of pre-cultural gregariousness is entirely unfounded; until a radical difference in nature is shown between human sociability, which is a cultural achievement, and animal gregariousness, which is an innate endowment, it is futile to show how social organization develops out of early kinship groups. Instead of having to face the 'herd instinct' at every turn of our argument and show its inadequacy then and there, it is best to deal with this mistaken point of view from the outset.

It is idle, I believe, to consider the purely zoological question whether our pre-human ancestors lived in big herds and were endowed with the necessary innate tendencies which allow animals to co-operate in herds, or whether they lived in single families. The question we have to answer is whether any forms of human organization can be derived from any animal types of herding; that is whether organized behaviour can be traced back to any forms of animal gregariousness or 'herd-instinct'.

from the lower classes in a society; the plurality of wives allows a chief to obtain economic and political advantages and thus provides a basis for distinctions of rank.

Let us first consider animal gregariousness. It is a fact that a number of animal species are so constituted that they have to lead their life in more or less numerous groups, and that they solve the main problems of their existence by innate forms of co-operation. Can we say with regard to such animal species that they possess a specific 'herd' or 'gregarious' instinct? All competent definitions of instinct agree that it must mean a *fixed pattern of behaviour*, associated with certain *anatomical mechanisms* correlated to *organic needs* and showing a general *uniformity throughout the species*. The various specific methods by which animals carry on the process of search for food, of nutrition; the series of instincts which constitute mating, the rearing and education of offspring; the working of the various locomotive arrangements; the functioning of primitive defensive and offensive mechanisms,— these constitute instincts. In each we can correlate the instinct with an anatomical apparatus, with a physiological mechanism and a specific aim in the vast biological process of individual and racial existence. Throughout the species each individual will behave in an identical manner, provided that the conditions of its organism and the external circumstances are present to release the instinct.

What about gregariousness? It is interesting to note that we find the division of functions, the co-ordination of activities, and the general integration of collective life most pronounced among relatively low forms of animal life such as the insects, and also, perhaps, coral colonies. (Compare the writer's article on 'Instincts and Culture' in *Nature*, July 19, 1924.) But neither with the social insects nor with gregarious mammals do we find a specific anatomical outfit subserving any specific act of 'herding'. The collective behaviour of animals subserves all processes, it envelops all instincts, but it is not a specific instinct. It might be called an innate component, a general modification of all instincts which makes the animals of the species co-operate in most vital affairs. It is important to note that in all the collective behaviour of animals co-operation is governed by innate adaptations and not by anything which could be called social organization in the sense in which we apply this word to humanity. This I have established more fully in the article mentioned above.

Thus man could not have inherited a gregarious instinct, which no animal possesses, but only a diffused 'gregariousness'. This would obviously mean that man has a general tendency to carry out certain adaptations by collective rather than individual behaviour, an assumption which would not help us very much in any concrete anthropological problem. Yet even the assumption of a tendency towards gregariousness can be shown to be completely erroneous. For is there any tendency in man to carry out all important acts in common; or even any well-defined type of activity 'gregariously'? He is capable indeed of developing his powers of co-operation indefinitely, of harnessing increasing numbers of his fellow creatures to one cultural task. But whatever type of activity be considered, man is also capable of carrying on his work in isolation if the conditions and the type of culture demand it. In the processes connected with nutrition and the satisfaction of bodily wants we can find every activity: food gathering, fishing, agriculture, performed either in groups or alone, by collective labour as well as by individual effort. In carrying out the propagation of the race man is capable of developing collective forms of sexual competition, of group licence side by side with strictly individual forms of courtship. The collective tending of offspring, found at least among insects, has no parallel in human societies, where we see individual parenthood devoted to the care of individual children. Again, while many ceremonies of religion and magic are performed in common, individual initiation rites, solitary experiences and personal revelation play as great a part in religion as do collective forms of worship. There is no more trace of gregarious tendencies in the domain of the *sacred* than in any other type of human culture.[1] Thus scrutiny of cultural activities would reveal no gregarious tendencies of any sort. As a matter of fact, the further we go back the more the individual character predominates, at least in economic work. It never becomes quite solitary, however, and the stage of 'individual search for food' postulated by certain economists seems to me to be a fiction: even at low levels organized activities run always side by side with individual effort. But there is no doubt that as culture advances individual activities gradually disappear from the economic field and are replaced by collective production on an enormous scale. We would have then a case of an 'instinct' increasing with culture, which, as can be easily seen, is a *reductio ad absurdum*!

Another way of approaching the question of the so-called 'herd instinct' would be to examine the nature of the bonds which unite men into social groups. These bonds, whether political, legal, linguistic, or customary are one and all of an acquired character; in fact, it can be easily seen that there is no innate element in them at all. Take the bonds of speech which unite groups of people at all levels of culture and sharply distinguish them from those with whom it is impossible to communicate by word of mouth. Language is an entirely acquired bodily habit. It is not based on any innate apparatus, it is completely dependent upon the culture and the tradition of a tribe, that is upon elements which vary within the same species, and so cannot be specifically innate. It is clear, moreover, that no 'language instinct' could have been inherited from our animal ancestors, who never communicated by a symbolic conventional code.

Whatever form of organized co-operation we take, we see after a brief scrutiny that it is based on cultural artefacts and governed by conventional norms. In the economic activities,

---

[1] This has been worked out in detail by myself in another publication, 'Magic, Religion and Science' in *Science, Religion and Reality*, collected essays by various authors, edited by J. Needham, 1925.

man uses tools and proceeds according to traditional methods. The social bonds which unite economic co-operative groups are therefore based upon a completely cultural framework. The same refers to an organization for purposes of war, of religious ceremonial, of the enforcement of justice. Nature could not have endowed human beings with specific responses towards arte- facts, traditional codes, symbolic sounds, for the simple reason that all these objects lie outside the domain of nature. The forms and forces of social organization are imposed upon a human community by culture and not by nature. There cannot be any innate tendency to run a locomotive or to use a machine gun, simply because these implements cannot have been anticipated by the natural conditions under which the human species has been biologically fashioned.

In all his organized behaviour man is always governed by those elements which are outside any natural endowment. Psy- chologically, human organization is based upon sentiments, that is complex built-up attitudes and not innate tendencies. Technic- ally, human association is always correlated with artefacts, with tools, implements, weapons, material contrivances all of which extend beyond man's natural anatomical equipment. Human sociality is always a combination, a dove-tailing of legal, polit- ical, and cultural functions. It is not a mere identity of the emo- tional impulse, not a similarity of response to the same stimulus, but an acquired habit dependent upon the existence of an arti- ficial set of conditions. All this will become clearer after our subsequent discussion of the formation of social bonds out of innate tendencies within the family.

To sum up, we can say that man obviously has to behave in common and that his organized behaviour is one of the corner- stones of culture. But while collective behaviour in animals is due to innate equipment, in man it is always a gradually built-up habit. Human *sociality* increases with culture, while if it had been mere *gregariousness* it should decrease or, at least, remain constant.

The fact is that the essential foundation of culture lies in a deep modification of innate endowment in which most instincts disappear, and are replaced by plastic though directed tendencies which can be moulded into cultural responses. The social integration of these responses is an important part of the process, but this integration is possible through the general plasticity of instincts and not through any specific gregarious tendency!

We may thus conclude that no type of human organization can be traced back to gregarious tendencies, still less to a specific 'herd instinct'. We shall be able to show that the necessary correlate of this principle is that the family is the only type of grouping which man takes over from the animal. In the process of transmission, however, this unit changes fundamentally with regard to its nature and constitution, though its form remains remarkably unaltered. The group of parents and children, the permanence of the maternal tie, the relation of father to his offspring, show remarkable analogies throughout human culture and in the world of higher animals. But as the family passes under the control of cultural elements, the instincts which have exclusively regulated it among pre-human apes become transformed into something which did not exist before man: I mean the cultural bonds of social organization. We have now to enquire into this transformation of instinctive responses into cultural behaviour.

3

RUT AND MATING IN
ANIMAL AND MAN

Let us compare the chain of linked instinctive responses which in animals constitute courtship, marriage and family with the corresponding human institutions. Let us, point after point, go over each link in the love-making and family life of anthropoid apes and ascertain what in human beings corresponds to each.

Among apes the courtship begins with a change in the female organism, determined by physiological factors and automatically releasing the sexual response in the male.[1] The male then

---

[1] In this context I should like to refer the reader to Havelock Ellis's *Studies in the Psychology of Sex* (six vols). In that work the biological nature of the regulation of the sexual instinct under culture is never lost sight of, and the parallel between animal and human societies is used as an important principle of explanation. For an interesting comment upon Darwin's Theory of Sexual Selection, see vol. iii, p. 22 *seqq.* (1919 ed.). In this volume the reader will also find a general criticism of the various theories of the sex impulse. In volume iv, Sexual Selection in man is discussed; volume vi deals with the sociological aspect of the problem.

proceeds to court according to the selective type of wooing which prevails in a given species. In this all the individuals who are within the range of influence take part, because they are irresistibly attracted by the condition of the female. Rut provides opportunities for display on the part of the males and for selection on the part of the female. All the factors which define animal behaviour at this stage are common to all individuals of the species. They work with such uniformity that for each animal species one set of data and only one has to be given by the zoologist, while, on the other hand, they vary considerably from one species to another, so that for each species a new description is necessary. But within the species the variations, whether individual or otherwise, are so small and irrelevant that the zoologist ignores them and is fully justified in doing so.

Could an anthropologist provide such a formula for the mechanism of courtship and mating in the human species? Obviously not. It is sufficient to open any book referring to the sexual life of humanity, whether it be the classical works of Havelock Ellis, Westermarck, and Frazer or the excellent description in Crawley's *Mystic Rose*, to find that there are innumerable forms of courtship and marriage, that seasons of love-making are different, that types of wooing and winning vary with each culture. To the zoologist the species is the unit, to the anthropologist the unit is the culture. In other words, the zoologist deals with specific instinctive behaviour, the anthropologist with a culturally fashioned habit-response.

Let us examine this in greater detail. In the first place we see that in man there is no season of rut, which means that man is ready to make love at any time and woman to respond to him—a condition which, as we all know, does not simplify human intercourse. There is nothing in man which acts with the same sharp determination as does the onset of ovulation in any mammalian female. Does this mean, however, that there is anything approaching indiscriminate mating in any human society? We know that even in the most licentious cultures nothing like 'promiscuity' exists or could ever have existed. In every human culture we find, first of all, systems of well-defined taboos which rigidly separate a number of people of opposite sexes and exclude whole categories of potential partners. The most important of these taboos completely excludes from mating those people who are normally and naturally in contact, that is, the members of the same family, parents from children, and brothers from sisters. As an extension of this, we find in a number of primitive societies a wider prohibition of sex intercourse which debars whole groups of people from any sex relations. This is the law of exogamy. Next in importance to the taboo of incest is the prohibition of adultery. While the first serves to guard the family, the second serves for the protection of marriage.

But culture does not exercise a merely negative influence upon the sexual impulse. In each community we find also inducements to courtship and to amorous interest besides the prohibitions and exclusions. The various festive seasons, times of dancing and personal display, periods when food is lavishly consumed and stimulants used, are as a rule also the signal for erotic pursuits. At such seasons large numbers of men and women congregate and young men are brought in contact with girls from beyond the circle of the family and of the local group. Very often some of the usual restraints are lifted and boys and girls are allowed to meet unhampered and uncontrolled. Indeed, such seasons naturally encourage courtship by means of the stimulants, the artistic pursuits, and the festive mood.[1]

Thus the signal for courtship, the release of the process of mating, is given not by a mere bodily change but by a

---

[1] Havelock Ellis has given a wealth of data on the seasonal mating in animal and man in the essay on *Sexual Periodicity*, vol. i (1910 ed.), especially see pp. 122 seqq. Compare also Westermarck's *History of Human Marriage*, vol. i, ch. ii.

combination of cultural influences. In the last instance these influences obviously act upon the human body and stimulate innate reactions in that they provide physical proximity, mental atmosphere, and appropriate suggestions; unless the organism were ready to respond sexually no cultural influences could make man mate. But, instead of an automatic physiological mechanism, we have a complicated arrangement into which artificial elements have been largely introduced. Two points, therefore, must be noted: there is no purely biological release mechanism in man, but instead there is a combined psychological and physiological process determined in its temporal, spatial, and formal nature by cultural tradition; associated with it and supplementing it is a system of cultural taboos which limit considerably the working of the sexual impulse.

Let us inquire now what is the biological value of rut for an animal species and what are the consequences for man of its absence. In all animal species mating has to be selective, i.e. there must be opportunities for comparison and for choice with either sex. Both male and female must have a chance to display his or her charms, to exercise attractions, to compete for the chosen one. Colour, voice, physical strength, cunning and agility in combat—each a symptom of bodily vigour and organic perfection—determine the choice. Mating by choice, again, is an indispensable counterpart of natural selection, for without some arrangement for selective mating the species would degenerate. This necessity increases as we ascend the scale of organic evolution; in the lowest animals there is not even the need for pairing. It is clear, therefore, that in the highest animal, man, the need for selective mating cannot have disappeared. In fact, the opposite assumption, that it is most stringent, is more likely to be true.

Rut, however, supplies the animal not only with the opportunities for selection. It also definitely circumscribes and delimits sexual interest. Outside the rutting season the sexual interest is in abeyance and the competition and strife as well as the overpowering absorption in sex are eliminated from the ordinary life of an animal species. Considering the great danger from outside enemies and the disruptive forces within, which are associated with courtship, the elimination of the sex interest from normal times and its concentration on a definite short period is of great importance for the survival of animal species.

In the light of all this, what does the absence of rut in man really signify? The sexual impulse is not confined to any season, not conditioned by any bodily process, and as far as mere physiological forces are concerned, it is there to affect at any moment the life of man and woman. It is ready to upset all other interests at all times; left to itself it tends constantly to work upon and loosen all existing bonds. This impulse, absorbing and pervading as it is, would thus interfere with all normal occupations of man, would destroy any budding form of association, would create chaos from within and would invite dangers from without. As we know, this is not a mere phantasy; the sex impulse has been the source of most trouble from Adam and Eve onwards. It is the cause of most tragedies, whether we meet them in present day actualities, in past history, in myth or in literary production. And yet the very fact of conflict shows that there exist some forces which control the sexual impulse; it proves that man does not surrender to his insatiable appetites; that he creates barriers and imposes taboos which become as powerful as the very forces of destiny.

It is important to note that these barriers and mechanisms which regulate sex under culture are different from the animal safeguards in the state of nature. With the animal instinctive endowment and physiological change throw male and female into a situation out of which they have to extricate themselves by the simple play of natural impulses. With man the control comes, as we know, from culture and tradition. In each society we find rules which make it impossible for men and women to yield freely to the impulse. How these taboos arise, by what forces they work, we shall see presently. For the moment it is enough to realize clearly that a social taboo does not derive its force from instinct, but that instead it always has to work against some innate impulse. In this we see plainly the difference between human endowment and animal instinct. While man is ready to respond sexually at any moment, he also submits to an artificially imposed check upon this response. Again, while there is no natural bodily process which definitely releases active sexual interest between male and female, a number of inducements towards courtship guide and bring out the impulse.

We can now formulate more precisely what we mean by the plasticity of instincts. The modes of behaviour associated with sex interest are determined in man only as regards their ends; man must mate selectively, he cannot mate promiscuously. On the other hand, the release of the impulse, the inducement to courtship, the motives for a definite selection are dictated by cultural arrangements. These arrangements have to follow certain lines parallel to the lines of natural endowment in the animal. There must be an element of selection, there must be safeguards for exclusiveness, above all there must be taboos which prevent sex from constantly interfering in ordinary life.

The plasticity of instincts in man is defined by the absence of physiological changes, of automatic release of a biologically determined cause of courtship. It is associated with the effective determination of sexual behaviour by cultural elements. Man is endowed with sexual tendencies but these have to be moulded in addition by systems of cultural rules which vary from one society to another. We shall be able to see with greater precision in the course of our present inquiry how far these norms can differ from each other and diverge from the fundamental animal pattern.

4

MARITAL RELATIONS

Let us follow the universal romance of life and look into its next stage. And let us examine the bonds of marriage into which lead the two parallel paths of man and animal, of eolithic cave-dweller and of super-simian ape. Of what does marriage really consist in animals, especially in apes? Mating occurs as the culminating act of courtship and with this the female conceives. With impregnation the rut is over and with its end there ceases the sexual attractiveness of the female to other males. But this is not the case with the male who has won her, whom she has chosen and to whom she has surrendered. It is difficult to affirm from the data at our disposal whether in the state of nature the higher apes still continue to mate sexually after impregnation. The fact, however, that the female ceases to be attractive to other males while her mate remains attached to her constitutes the bond of animal marriage. The specific response of both male and female to the new situation; their mutual attachment; the tendency of the male to remain with his consort, to guard her, to assist her, and to protect and nourish her—these are the innate elements of which animal marriage is made up. The new phase of life therefore consists of a new type of behaviour; it is dominated by a new link in the chain of instincts. This new link might appropriately be called the matrimonial response in contrast to the sexual impulse. The animal union is based neither upon the uncontrollable passion of rut nor on the sexual jealousy of the male nor on any claims of general appropriation on the part of the male. It is based on a special innate tendency.

When we pass to human society the nature of matrimonial bonds is found to be entirely different. The act of sexual union, in the first place, does not constitute marriage. A special form of ceremonial sanction is necessary and this type of social act differs from the taboos and inducements of which we spoke in the previous chapter. We have here a special creative act of culture, a sanction or hallmark which establishes a new relation between two individuals. This relationship possesses a force derived not from instincts but from sociological pressure. The new tie is something over and above the biological bond. As long as this creative act has not been performed, as long as marriage has not been concluded in its cultural forms, a man and a woman can mate and cohabit as long and as often as they like, and their relation remains something essentially different from a socially sanctioned marriage. Their tie, since there is no innate matrimonial arrangement in man, is not biologically safeguarded. Nor is it, since society has not established it, enforced by cultural sanction. As a matter of fact, in every human society a man and a woman who attempt to behave as if they were married without obtaining the appropriate social sanction are made to suffer more or less severe penalties.

A new force, therefore, a new element, comes into play supplementing the mere instinctive regulation of animals: the actual interference of society. And it need hardly be added that once this sanction has been obtained, once two people have been married, they not only may but must fulfill the numerous obligations, physiological, economic, religious, and domestic which are involved in this human relationship. As we have seen, the conclusion of a human marriage is not the consequence of a mere instinctive drive but of complex cultural inducements. But after matrimony has been sociologically sealed and hall-marked, a number of duties, ties, and reciprocities are imposed, backed up by legal, religious, and moral sanctions. In human societies such a relationship can usually be dissolved and re-entered with another partner but this process is never easy to carry out, and in some cultures the price of divorce makes it almost prohibitive.

Here we see clearly the difference between instinctive regulation on the one hand and cultural determinism on the other. While in animals marriage is induced by selective courtship, concluded by the mere act of impregnation, and maintained by the forces of the innate matrimonial attachment, in man it is induced by cultural elements, concluded by sociological sanction and maintained by the various systems of social pressure. And yet here again it is not difficult to perceive that the cultural apparatus works very much in the same direction as natural instincts and that it attains the same ends though the mechanism entirely differs. In the higher animals marriage is necessary because the longer the pregnancy, the more helpless the pregnant female and the new-born infant and the more necessary it is for them to have the protection of the male. The innately determined bond of matrimonial affection by which the male responds to the pregnancy of his chosen mate fulfills this need of the species, and is, in fact, indispensable for its continuity.

In man this need for an affectionate and interested protector of pregnancy still remains. That the innate mechanism has disappeared we know from the fact that in most societies on a low as well as on a high level of culture the male refuses to take any responsibility for his offspring unless compelled to do so by society, which enforces the contract of marriage. But each culture develops certain forces and there exist certain arrangements which play the same part as the instinctive drives do in an animal species. The institution of marriage in its fundamental moral, legal and religious aspects must thus be regarded not as the direct outgrowth of the matrimonial tendency in animals but as its cultural substitute. This institution imposes upon man and woman a type of behaviour which corresponds as closely to the needs of the human species as the innate tendencies in animals correspond to theirs.

As we shall see, the most powerful means by which culture binds husband and wife to each other consists in the moulding and organizing of their emotions and in the shaping of their personal attitudes. This process we shall have opportunity to study more fully, and in it we shall find the essential differences between animal and human bonds. While in animals we find a chain of linked instincts succeeding each other and replacing each other, human behaviour is defined by a fully organized emotional attitude, a *sentiment*, as it is technically called in psychology. While in the animal we have a series of physiological moments, events happening within the organism, each of which determines an innate response, in man we have a continuously developing system of emotions. From the first meeting of the two prospective lovers, through gradual infatuation and the growth of associated interests and affections, we can follow a developing and increasingly richer system of emotions in which continuity and consistency are the condition of a happy and harmonious relationship. Into this complex attitude there enter, besides innate responses, social elements, such as moral rules, economic expectations and spiritual interests. The latter stages of matrimonial affection are powerfully determined by the course of courtship. On the other hand, courtship and the personal interest of two prospective lovers is coloured by the possibilities of future matrimony and by its advantages. In the anticipatory elements, in which the future responses are brought to bear upon present arrangements; in the influence of memories and experiences; in the constant adjustment of past, present, and future, we see why human relationship presents a continuous and homogeneous growth instead of the series of clearly differentiated stages which we find in the animal.

In all this, again, we meet the same plasticity of instincts already noticed in the earlier stages, and we see that though the mechanisms under culture differ considerably from physiological arrangements, the general forms into which society moulds human matrimonial rules follow clearly the lines dictated by natural selection to animal species.

5

PARENTAL LOVE

Courtship, mating, and pregnancy lead in animal and man to the same end: the birth of the offspring. To this event there is also a similar mental response in pre-human species as well as in woman and man under culture. In fact at first sight the act of birth might be quoted as the one organic event in which man does not differ at all from the animal. Maternity, indeed, is usually regarded as the one relationship which is bodily carried over from the simian to the human status; which is defined biologically and not culturally. This view, however, is not correct. Human maternity is a relationship determined to a considerable degree by cultural factors. Human paternity, on the other hand, which appears at first as almost completely lacking in biological foundation, can be shown to be deeply rooted in natural endowment and organic need. Thus here again we are forced to compare minutely the animal with the human family, to state the similarities as well as the differences.

With the animal, birth changes the relationship between the two mates. A new member has arrived into the family. The mother responds to it immediately. She licks the offspring, watches it constantly, warms it with her body, and feeds it with her breasts. The early maternal cares imply certain anatomical arrangements such as the pouches in the marsupials and the milk-glands in the mammals. There comes a response in the mother to the appearance of the offspring. There is also a response in the young—it is, in fact, perhaps the most unquestionable type of instinctive activity.

The human mother is endowed with similar anatomical equipment and, in her body, conception, pregnancy, and child-birth entail a series of changes analogous to the gestation of any other mammal. When the child is born the bodily status which constitutes animal motherhood is to be found also in the human mother. Her breasts swollen with milk invite the child to suck with an impulse as elementary and powerful as the infant's hunger and thirst. The needs of the child for a warm, comfortable, and safe place dovetail into the extremely strong, passionate desire of the mother to clasp the infant. They are correlated to her tenderness and solicitude for the child's welfare.

Yet in no human society, however high or low it might be in culture, is maternity simply a matter of biological endowment or of innate impulses. Cultural influences analogous to those we found determining relations between lovers and imposing obligations between consorts, are at work even in moulding the relation of the mother to the child. From the moment of conception this relation becomes a concern of the community. The mother has to observe taboos, she follows certain customs and submits to ritual proceedings. In higher societies these are largely but not completely replaced by hygienic and moral rules; in lower they belong to the domain of magic and religion. But all such customs and precepts aim at the welfare of the unborn child. For its sake the mother has to undergo ceremonial treatment, suffer privations and discomforts. Thus an obligation is imposed upon the prospective mother in anticipation of her future instinctive response. Her duties run ahead of her feelings, culture dictates and prepares her future attitude.

After birth the scheme of traditional relations is not less powerful and active. Ceremonies of purification, rules which isolate the mother and child from the rest of the community, baptismal rites and rites of the reception of the newborn infant into the tribe, create one and all a special bond between the two. Such customs exist both in patrilineal and matrilineal societies. In these latter there are, as a rule, even more elaborate arrangements and the mother is brought into yet closer contact with the child, not only at the outset but also at a later period.

Thus it can be said without exaggeration that culture in its traditional bidding duplicates the instinctive drive. More precisely it anticipates its rulings. At the same time, all cultural influences simply endorse, amplify, and specialize the natural tendencies, those which bid the mother tenderly to suckle, to protect, and to care for her offspring.

If we try to draw the parallel between the relation of father to child in animal and human societies, we find that it is easy to discover the cultural elements in humanity but difficult to find out what instinctive endowment could exist. As a matter of fact, in higher cultures at least the necessity for imposing the bond of marriage is practically and theoretically due to the fact that a father has to be made to look after his children. An illegitimate child has, as a rule, no chance of receiving the same care from its natural father as a legitimate one and the latter is cared for to a large extent because it is the father's duty. Does that mean that there are no innate paternal tendencies in man? It will be possible for us to show that the human father is, on the contrary, endowed with definite impulses—not sufficient to establish natural paternity, but powerful enough to serve as the raw material out of which custom is fashioned.

Let us first look at paternity among the higher mammals. We know that the male is indispensable there, because, owing to long pregnancy, lactation, and education of the young, the female and her offspring need a strong and interested protector. Correlated with this need we find what has been called in the previous chapter the matrimonial response. This response, which induces the male to look after the pregnant female, is not weakened by the act of birth, but, on the contrary, it becomes stronger and develops into a tendency on the part of the male to protect the whole family. The matrimonial attachment between the two partners has to be regarded biologically as an intermediate stage leading up to paternal attachment.

Turning now to human societies, we see that the need, far from abating, becomes even stronger. The pregnant and lactant woman is not less but more helpless than her simian sister, and this helplessness increases with culture. The children again need not only the ordinary cares of animal infancy, not merely suckling and tending, as well as the education of certain innate tendencies, but also such instruction in language, tradition, and handicraft as is indispensable even in the simplest human societies.

Can we therefore imagine that as humanity was passing from a state of nature into culture the fundamental tendency in the male, which under the new conditions was even more imperative, should be gradually lessened or be led to disappear? Such a state of affairs would run counter to all biological laws. It is, in fact, completely denied by all the facts observed in human societies. For, once a man is made to remain with his wife to guard her pregnancy, to observe the various duties which he usually has to fulfill at birth, there can be not the slightest doubt that his response to the offspring is that of impulsive interest and tender attachment.

Thus we see an interesting difference between the working of cultural and natural endowment. Culture—in the form of law, morals, and custom—forces the male into the position in which he has to submit to the natural situation, that is, he has to keep guard over the pregnant woman. It forces him also, through various means, to share in her anticipatory interest in the child. But once forced into this position, the male responds invariably with strong interests and positive feelings for the offspring.

And this brings us to a very interesting point. In all human societies—however they might differ in the patterns of sexual morality, in the knowledge of embryology, and in their types of courtship—there is universally found what might be called the rule of legitimacy. By this I mean that in all human societies a girl is bidden to be married before she becomes pregnant. Pregnancy and childbirth on the part of an unmarried young woman are invariably regarded as a disgrace.[1] Such is the case in the very free communities of Melanesia described in this essay. Such is the case in all human societies concerning which we have any information. I know of no single instance in anthropological literature of a community where illegitimate children, that is children of unmarried girls, would enjoy the same social treatment and have the same social status as legitimate ones.

The universal postulate of legitimacy has a great sociological

---

[1] Westermarck in the *History of Human Marriage*, 1921, vol. i, pp. 138–57, cites approximately 100 cases of primitive peoples who are characterized by their pre-nuptial chastity. But many of the statements quoted do not afford very definite evidence of this fact. Thus to say of certain tribes that 'chastity is prized in man or woman' or that 'a good deal of value is laid upon the virginity of the bride' is not to give proof of lack of pre-marital intercourse. What, however, is of extreme importance in this computation of evidence from our point of view is that the only thing it definitely indicates is the universality of the *postulate of legitimacy*. Thus twenty-five of the cases quoted refer, not to chastity, but to the prohibition of an unmarried girl being with child. Further, more than a score of others indicate, not the absence of illicit sexual relations, but that when they occur, censure, or punishment, or a fine, or compulsion of the two parties to marry, according to the tribe, follows discovery. In fact though the total evidence is not conclusive as regards chastity it does prove that the postulate of legitimacy is of extremely wide prevalence. The two problems should be kept distinct, from the point of view of our argument.

significance, which is not yet sufficiently acknowledged. It means that in all human societies moral tradition and law decree that the group consisting of a woman and her offspring is not a sociologically complete unit. The ruling of culture runs here again on entirely the same lines as natural endowment; it declares that the human family must consist of the male as well as the female.

And in this culture finds a ready response in the emotional attitudes of the male. The father at all stages of culture is interested in his children, and this interest, no matter how it might be rationalized in certain patrilineal societies, is exactly the same in matrilineal societies where the child is neither an heir nor successor to his father nor even usually regarded as the offspring of his body.[1] And even when, as in a polyandrous society, there is no possibility at all for any knowledge and interest in the matter of who might be the begetter, the one who is selected to act as the father responds emotionally to this call.

It would be interesting to inquire in what way we could imagine the working of the instinctive tendency of fatherhood. With the mother the response is plainly determined by the bodily facts. It is the child whom she has created in her womb that she is going to love and be interested in. With the man there can be no such correlation between the seminal cell which fertilizes the female ovum on the one hand and the sentimental attitude on the other. It seems to me that the only factors which determine the sentimental attitude in the male parent are connected with the life led together with the mother during her pregnancy. If this is correct, we see how the dictates of culture are necessary in order to stimulate and organize emotional attitudes in man and how innate endowment is indispensable to culture. Social forces alone could not impose so many duties on the male, nor

---

[1] Compare the writer's *The Father in Primitive Psychology*, 'Psyche Miniatures,' 1927.

without a strong biological endowment could he carry them out with such spontaneous emotional response.

The cultural elements which enter into the father-to-child relationship are closely parallel to those which determine maternity. The father usually has a share in the mother's taboos, or, at least, he has to maintain some others side by side with her. A special type of prohibition which is definitely associated with the welfare of the child is the taboo on sexual intercourse with a pregnant wife. At birth there are again duties for the father to perform. The most famous of these is the *couvade*, a custom in which the husband has to take over the symptoms of post-natal illness and disability while the wife goes about the ordinary business of life. But though this is the most extreme form of affirmation of paternity, some analogous arrangement, by which the man shares in certain post-natal burdens of his wife, or, at least, has to carry on actions in sympathy with her, exist in all societies. It is not difficult to place this type of custom in our scheme. Even the apparently absurd idea of the *couvade* presents to us a deep meaning and a necessary function. If it is of high biological value for the human family to consist of both father and mother; if traditional customs and rules are there to establish a social situation of close moral proximity between father and child; if all such customs aim at drawing the man's attention to his offspring, then the *couvade* which makes man simulate the birth-pangs and the illness of maternity is of great value and provides the necessary stimulus and expression for paternal tendencies. The *couvade* and all the customs of its type serve to accentuate the principle of legitimacy, the child's need of a father.

In all this we have again the two sides of the question. Instincts alone never determine human behaviour. Rigid instincts which would prevent man's adaptation to any new set of conditions are useless to the human species. The plasticity of instinctive tendencies is the condition of cultural advance. But the tendencies are there and cannot be developed arbitrarily. Although the character of the maternal relation is determined by culture; although the obligations are imposed from outside by tradition, they all correspond to the natural tendency, for they all emphasize the closeness of the bond between father and child, they isolate them and make them dependent upon each other. It is important to note that many of these social relations are anticipatory: they prepare the father for his future feelings, they dictate to him beforehand certain responses, which he will later develop.

Paternity we have seen cannot be regarded as a merely social arrangement. Social elements simply place man into a situation in which he can respond emotionally, and they dictate to him a series of actions by which the paternal tendencies can find their expression. Thus, while we find that maternity is social as well as biological, we must affirm that paternity is determined also by biological elements, that therefore in its make-up it is closely analogous to the maternal bond. In all this culture emphasizes rather than overrides the natural tendencies. It re-makes, with other elements, the family into the same pattern as we find in nature. Culture refuses to run riot.

6

THE PERSISTENCE OF FAMILY TIES IN MAN

The family life of mammals always lasts beyond the birth of the offspring and the higher the species the longer both parents have to look after their progeny. The gradual ripening of the child needs more protracted care and training on the part of both father and mother, and these have to remain united to look after the little ones. But in no animal species does the family last for life. As soon as the children are independent they leave the parents. This is in keeping with the essential needs of the species, for any association, with its corresponding ties becomes a burden to animals unless it has some specific function to fulfill.

With man, however, new elements enter. Apart from the tender cares dictated by nature and endorsed by custom and tradition, there enters the element of cultural education. Not only is there a need of training instincts into full development, as in the animal instruction in food-gathering and specific movements, there is also the necessity of developing a number of cultural habits as indispensable to man as instincts are to animals. Man has to teach his children manual skill and knowledge in arts and crafts; language and the traditions of moral culture; the manners and customs which make up social organization.

In all this there is the need of special co-operation between the two generations, the older which hands on and the younger which takes over tradition. And here we see the family again as the very workshop of cultural development, for continuity of tradition, especially at the lowest levels of development, is the most vital condition of human culture and this continuity depends upon the organization of the family. It is important to insist that with the human family this function, the maintenance of the continuity of tradition, is as important as the propagation of the race. For man could no more survive if he were deprived of culture than culture could survive without the human race to carry it on. Newer psychology teaches us, moreover, that the earliest steps of human training, those which happen within the family, are of an educational importance which has been completely overlooked by earlier students. But if the influence of the family is enormous at present, it must have been even greater at the beginnings of culture, where this institution was the only school of man and the education received was simple but had to be given with a vigour of outline and a strength of imperative not necessary at higher levels.

In this process of parental education by which the continuity of culture is maintained we see the most important form of division of functions in human society: that between giving the lead and taking it, between cultural superiority and inferiority. Teaching—the process of imparting technical information and moral values—requires a special form of co-operation. Not only must the parent have an interest in instructing the child, and the child an interest in being taught, but a special emotional setting is also necessary. There must be reverence, submission, and confidence on the one hand, tenderness, feeling of authority, and desire to guide on the other. Training cannot be done without some authority and prestige. The truths revealed, the examples given, the orders imposed will not reach their aim or command obedience unless they are backed up by those specific attitudes of tender subordination and loving authority which are characteristic of all sound parental relations to the child. These correlated attitudes are most difficult and most important in the relation between the son and the father. Owing to the vigour and initiative of the young and the conservative authority of the old male, there is a certain difficulty in the establishment of a permanent reverent attitude. The mother, as the nearest guardian and the most affectionate helpmate, usually finds no difficulties in the earlier stages of relation to children. In the relation between son and mother, however, which, if it is to continue harmonious, should remain one of submission, reverence, and subordination, there enter other disturbing elements at a later stage of life. To these, already known from the previous parts of this book, we shall presently have once more to return.

The mature animal departs naturally from its parents. In man the need for more enduring bonds is indisputable. The education of the children, first of all, binds them to the family for a long period beyond their maturity. But even the end of cultural education is not the final signal for dissolution. The contacts established for cultural training last longer, and they serve for the establishment of further social organization.

Even after a grown-up individual has left his parents and established a new household his relation to them remains active. In all primitive societies, without exception, the local community, the clan or the tribe, is organized by a gradual extension of family ties. The social nature of secret societies, totemic units and tribal groups is invariably based on courtship ideas, associated with local habitation by the principle of authority and rank, but with all this it is still definitely linked with the original family bond.[1]

It is in this actual and empirical relationship between all wider social groupings on the one hand and the family on the other that we have to register the fundamental importance of the latter. In primitive societies the individual does build up all his social ties upon the pattern of his relation to father and mother, to brother and sister. In this, again, anthropologists, psycho-analysts and psychologists are fully in agreement, putting on one side the fantastic theories of Morgan and some of his followers. Thus the endurance of family ties beyond maturity is the pattern of all social organization, and the condition of co-operation in all economic, religious, and magical matters. This conclusion we reached in a previous chapter, where we examined the alleged gregarious instinct and found that there is neither an instinct nor a tendency towards 'herding'. But if social bonds cannot be reduced to pre-human gregariousness, they must have been derived from the development of the only relationship which man has taken over from his animal ancestors: the relationship between husband and wife, between parents and children, between brothers and sisters, in short the relationship of the undivided family.

This being so, we see that the endurance of family bonds and the corresponding biological and cultural attitude is indispensable not only for the sake of the continuity of tradition but also for the sake of cultural co-operation. And in this fact we have to register what is perhaps the deepest change in the instinctive endowment of animal and man, for in human society the extension of family bonds beyond maturity does not follow the instinctive pattern to be found among animals. We can no longer

---

[1] I cannot document this point of view more extensively here. It will be developed in a work on *The Psychology of Kinship*, in preparation for the International Library of Psychology.

speak of plastic innate tendencies, for, since the family bonds extended beyond maturity do not exist in animals, they cannot be innate. Moreover the utility and function of life-long family bonds are conditioned by culture and not by biological needs. Parallel to this, we see that in animals there is no tendency to maintain the family beyond the stage of biological serviceability. In man, culture creates a new need, the need to continue close relations between parents and children for the whole life. On the one hand this need is conditioned by the transmission of culture from one generation to another; on the other by the need of life-long endurance of bonds which form the pattern and starting-point for all social organization. The family is the biological grouping to which all kinship is invariably referred and which determines by rules of descent and inheritance the social status of the offspring. As can be seen, this relation never becomes irrelevant to a man and has constantly to be kept alive. Culture, then, creates a new type of human bond for which there is no prototype in the animal kingdom. And as we shall see, in this very creative act, where culture steps beyond instinctive endowment and natural precedent, it also creates serious dangers for man. Two powerful temptations, the temptation of sex and that of rebellion, arise at the very moment of cultural emancipation from nature. Within the group which is responsible for the first steps in human progress there arise the two main perils of humanity: the tendency to incest and the revolt against authority.

7

THE PLASTICITY OF
HUMAN INSTINCTS

We shall proceed presently to discuss at some length the two
perils of incest and revolt, but first let us rapidly survey the gist
of the last few chapters in which the family among man and
animals has been compared. We found that in both the general
course of behaviour is paralleled in its external form. Thus there
exists a circumscribed courtship, usually limited in time, and
defined in its form both in human communities and in animal
species. Again, selective mating leads to an exclusive matrimonial life of which the monogamous marriage is a prevalent
type. Finally in animal and in man we found parenthood, imply-
ing the same kind of cares and obligations. In short, the forms of
behaviour and their functions are similar. The preservation
of species through selective mating, conjugal exclusiveness, and
parental care is the main aim of human institutions as well as of
animal instinctive arrangements.

Side by side with similarities we found conspicuous differences. These were not in the ends but in the means by which the

ends were reached. The mechanism by which the selection of mating is carried on, by which matrimonial relations are maintained and parental cares established, is in the animal entirely innate and based on anatomical endowment, physiological change, and instinctive response. The whole series shows the same pattern for all animals of the species. In man the mechanism is different. While there exists a general tendency to court, to mate, and to care for the offspring and while this tendency is as strong in man as in the animal, it is no longer clearly defined once and for all throughout the species. The landmark has disappeared and has been replaced by cultural limitations. The sexual impulse is permanently active, there is no rut nor any automatic disappearance of female attraction afterwards. There is no natural paternity, and even the maternal relations are not exclusively defined by innate responses. Instead of the precise instinctive determinants we have cultural elements which shape the innate tendencies. All this implies a deep change in the relation between instinct and physiological process and the modification of which they are capable. This change we have termed the 'plasticity of instincts'. The expression covers the set of facts described above in detail. They all show that in man the various physiological elements which release instinct have disappeared, while at the same time there appears a traditional training of the innate tendencies into cultural habit responses. These cultural mechanisms were analyzed concretely. They are the taboos which forbid incest and adultery; they are the cultural releases of the mating instinct; they are the moral and ideal norms as well as the practical inducements which keep husband and wife together—the legal sanction of the marriage tie; the dictates which shape and express parental tendencies. As we know, all these cultural co-determinants closely follow the general course imposed by nature on animal behaviour. In detail, however, the concrete forms of courtship, matrimony and parenthood vary with the culture and the forces by which they shape human behaviour are no longer mere instincts but habits into which man has been educated by tradition. The social sanction of law, the pressure of public opinion, the psychological sanction of religion and the direct inducements of reciprocity replace the automatic drives of the instincts.

Thus culture does not lead man into any direction divergent from the courses of nature. Man still has to court his prospective mate and she still has to choose and to yield to him. The two still have to keep to one another and be ready to receive the offspring and watch over them. The woman still has to bear and the man to remain with her as her guardian. Parents still have to tend and educate their children, and under culture they are as attached to them as under nature animals are to their offspring. But in all this an astounding variety of patterns replaces in human societies the one fixed type imposed by instinctive endowment upon all the individuals of a single animal species. The direct response of instinct is replaced by traditional norms. Custom, law, moral rule, ritual, and religious value enter into all the stages of love-making and parenthood. But the main line of their action invariably runs parallel to that of animal instincts. The chain of responses which regulate animal mating constitute a prototype of the gradual unfolding and ripening of man's cultural attitude. We must pass now to a more detailed comparison of the processes of animal instincts and human sentiments.

8

FROM INSTINCT
TO SENTIMENT

In the last chapter we summarized the salient points of our comparison between the constitution of the animal and the human family. Through the disappearance of the definite physiological landmarks, through the increasing cultural control in man there arises a complexity in the human response, a variety which at first seems to introduce nothing but chaos and disorder. This, however, is not really the case. In the first place we can see that the varying emotional adjustments of mating in the human family are simplified in one direction. The human bonds culminate on their sexual side in marriage, on their parental side in a life-long enduring family. In both cases the emotions centre around one definite object, whether this be the consort, the child, or the parent. Thus the exclusive dominance of one individual appears as the first characteristic in the growth of human emotional attitudes.

As a matter of fact we can see this tendency even as we ascend in the animal kingdom from the lower to the higher species.

Among the lower animals the male seed is often scattered broad-cast and the fertilizing of the female egg is left entirely to phys-ical agencies. The personal equation, selection and adjustment develop gradually and attain their fullest development among the highest animals.

In man, however, this tendency is translated and enforced by definite institutions. Mating, for instance, is defined by a number of sociological factors some of which exclude a number of females, while others indicate the suitable partners or stipulate definite unions. In certain forms of marriage the individual bond is completely established by social elements, such as infant betrothal or socially prearranged marriages. In any case, right through courtship, matrimonial relations and the care of the children, the two individuals gradually establish an exclusive personal tie. A number of interests of economic, sexual, legal, and religious nature are for each partner dominated by the per-sonality of the other. The legal and the religious sanction of marriage establishes, as we know, a lifelong, socially enforceable bond between the two. Thus in human relations the emotional adjustments are dominated by one object rather than by the situation of the moment. Within the same relationship the emo-tions and the type of drives and interests vary: they are usually one-sided and disconnected at the beginning of the courtship, they gradually ripen into a personal affection during that period, they are immensely enriched and complicated by the common life in marriage, even more so by the arrival of children. Yet throughout this variety of emotional adjustments the perman-ence of the object, its deep hold on the other individual's life constantly increases. The bond cannot be broken easily and the resistances are usually both psychological and social. Divorce in savage and civilized communities, for instance, or a rupture between parent and child is both a personal tragedy and a sociological mishap.

But though the emotions which enter into the human family

bond are constantly changing—though they depend upon circumstances—matrimonial love, for instance, entailing love and sorrow as well as joy, fear, and passionate inclinations—though they are always complex and never exclusively dominated by an instinct, yet they are by no means chaotic or disorganized, in fact they are arranged into definite systems. The general attitude of one consort to the other, of a parent to a child and vice versa is not in any way accidental. Each type of relationship must dispose of a number of emotional attitudes which subserve certain sociological ends, and each attitude gradually grows up according to a definite scheme through which the emotions are organized. Thus in the relations between the two consorts the sentiment begins with the gradual awakening of sexual passion. In culture, this, as we know, is never a merely instinctive moment. Various factors, such as self-interest, economic attraction, social advancement, modify the charm of a girl for a man or vice versa, in low levels of culture as well as in more highly developed civilizations. This interest once aroused, the passionate attitude has to be gradually built up by the traditional, customary course of courtship prevailing in a given society. No sooner has this attachment been built up, than the decision to enter marriage introduces a first contract, establishes a more or less sociologically defined relationship. Through this period a preparation for matrimonial ties takes place. The legal bond of marriage as a rule changes the relationship in which the sexual elements are still predominant into one of common life, and here the emotional attitudes have to become reorganized. It is important to note that the change from courtship to matrimony, which in all societies is the subject of proverbs and jokes, entails a definite and difficult readjustment of attitudes: while in the human relationship the sexual elements are not eliminated nor the memories of courtship effaced, entirely new interests and new emotions have to be incorporated. The new attitudes are built upon the foundation of the old and personal tolerance and patience in trying situations have to be formed at the expense of sexual attractiveness. The initial charms and the gratitude for the erotic pleasure of earlier life have a definite psychological value and form an integral part of the later feelings. We find in this an important element of human sentiments: the carrying over of previous memories into later stages. We shall presently analyze the relation of mother to child and father to son, and show there that the same system of gradual ripening and organizing the emotions takes place. There is always a dominant emotional attitude associated with the bodily relation. Between husband and wife sexual desire is indispensable, as well as an associated bond of personal attractiveness and compatibility of character. The sentimental elements of courtship, the passionate feelings of first possession must be incorporated into the calmer affection, allowing husband and wife to enjoy each other's company throughout the best part of their days. These elements must also be harmonized with the community of work and community of interest which unite the two into the joint managers of the household. It is a well-known fact that each transition between courtship and sexual cohabitation, between that stage and the fuller common life of later matrimony, between married life and parental life, constitutes a crisis full of difficulties, dangers and maladjustments. These are the points at which the attitude undergoes a special phase of reorganization.

The mechanism which we see at work in this process is based on a reaction between innate drives, human emotions and social factors. As we have seen, the organization of a society has economic, social, and religious ideals to impress upon the sexual inclination of men and women. These exclude certain mates by rules of exogamy, of caste division, or of mental training. They surround others by a spurious halo of economic attractiveness, of high rank, or superior social status. In the relation between parents and children also tradition dictates certain attitudes which even anticipate the appearance of the objects to which they pertain. The action of the sociological mechanisms is specially important when we see it at work in the growing mentality of the young. Education, especially in simpler societies, does not take place by the explicit inculcation of sociological, moral and intellectual principles, but rather by the influence of the surrounding cultural environment on the ripening mind. Thus the child learns the principles of caste, rank or clan division by the concrete avoidances, preferences and submissions into which he is being trained by practical measures. A certain ideal is thus impressed upon the mind and by the time the sexual interest begins to act the taboos and the inducements, the forms of correct courtship, the ideals of desirable matrimony are framed in his mind. It is imperative to realize that this moulding and gradual inculcation of ideals is not done by any mysterious atmosphere, but by a number of well-defined, concrete influences. If we cast back to the previous ideas of this book and follow the life of an individual in peasant Europe or savage Melanesia we can see how the child within the parental home is educated by the rebuke of the parents, by the public opinion of the elder men, by the feeling of shame and discomfort aroused by the reactions of them to certain types of his conduct. Thus the categories of decent and indecent are created, the avoidance of forbidden relationships, the encouragement towards certain other groups and the subtler tones of feeling toward the mother, father, maternal uncle, sister and brother. As a final and most powerful framework of such a system of cultural values, we have to note the material arrangements of habitation, settlement, and household chattels. Thus in Melanesia the individual family house, the bachelors' quarters, the arrangements of patrilocal marriage and of matrilineal rights are all associated, on the one hand with the structure of villages, houses, and the nature of territorial divisions, and on the other with injunctions, taboos, moral laws, and various shades of feeling. From this we can see that man gradually expresses his emotional attitudes in legal, social, and material arrangements and that these again react upon his conduct by moulding the development of his behaviour and outlook. Man shapes his surroundings according to his cultural attitudes, and his secondary environment again produces the typical cultural sentiments.

And this brings us to a very important point which will allow us to see why in humanity instinct had to become plastic and innate responses have to be transformed into attitudes or sentiments.

Culture depends directly upon the degree to which the human emotions can be trained, adjusted, and organized into complex and plastic systems. In its ultimate efficiency culture gives man mastery over his surroundings by the development of mechanical things, weapons, means of transportation and measures for protection against weather and climate. These, however, can only be used if side by side with the apparatus there is also transmitted the traditional knowledge and art of using it. The human adjustments to the material outfit have to be learned anew by each generation. Now this learning, the tradition of knowledge, is not a process which can be carried on by sheer reason nor by mere instinctive endowment. The transmission of knowledge from one generation to another entails hardships, efforts and an inexhaustible fund of patience and love felt by the older generation for the younger. This emotional outfit, again, is only partly based on the endowment, for all the cultural actions which it dominates are artificial, non-specific, and therefore not provided with innate drives. The continuity of social tradition, in other words, entails a personal emotional relation in which a number of responses have to be trained and developed into complex attitudes. The extent to which parents can be taxed with burdens of cultural education depends upon the capacity of the human character for adaptation to cultural and social responses. Thus in one of its aspects culture is directly dependent upon the plasticity of innate endowment.

But the relation of man to culture consists not only in transmission of tradition from one individual to another; culture even in its simplest forms cannot be handled except by co-operation. As we have seen, it is the lengthening of the ties within the family beyond strict biological maturity which allows on the one hand of cultural education and on the other of work in common, that is, co-operation. The animal family of course has also a rudimentary division of functions, consisting mainly in the provision of food by the male during certain stages of maternal care and later on in the protection and nutrition afforded by father and mother. In animal species, however, both the nutritive adjustment to environment and the scheme of economic division of functions are rigid. Man is allowed through culture to adapt himself to a very wide range of economic environment and this he controls, not by rigid instincts, but by the capacity for developing special technique, special economic organization, and adjusting himself to a special form of diet. But side by side with this merely technical aspect there must also go an appropriate division of function and a suitable type of co-operation. This obviously entails various emotional adjustments under various environmental conditions. The economic duties of husband and wife differ. Thus in an arctic environment the main burden of providing food falls on the man; among the more primitive agricultural peoples the woman has the greater share of providing for the household. With the economic division of functions there are associated religious, legal and moral distinctions which dovetail into economic work. The charm of social prestige, the value of the consort as practical helpmate, the ideal of moral or religious nature, all these considerably colour the relationship. It is its variety and the possibility of adjusting such relations as the conjugal one and the parental one which allow the family to adapt itself to varying conditions of practical co-operation and this latter to become adapted to the material outfit of culture and the natural environment. How far we can trace concretely these dependencies and correlations is beside the point in our present argument. I wish to emphasize here the fact that only plastic social ties and adjustable systems of emotion can function in an animal species which is capable of developing a secondary environment and thus adjusting itself to the difficult outer conditions of life.

Through all this we can see that although the basis of human family relations is instinctive, yet the more they can be moulded by experience and by education, the more cultural and traditional elements these ties can accept, the more suitable will they be for a varied and complex division of functions.

What has been said here with reference to the family refers obviously also to other social ties. But in these, contrary to what is the case with the bonds of the family, the instinctive element is almost negligible. The great theoretical importance of marriage and of the family runs parallel with the great practical importance of these institutions for humanity. Not only is the family the link between biological cohesion and social cohesion, it is also the pattern on which all wider relations are based. The further sociologists and anthropologists can work out the theory of sentiments, of their formation under cultural conditions and of their correlation with social organization, the nearer shall we move towards a correct understanding of primitive sociology. Incidentally I think that an exhaustive description of primitive family life, primitive courtship customs and clan organization will rule out from sociology such words as 'group instinct', 'consciousness of kind', 'group mind', and similar sociological verbal panaceas.

To those acquainted with modern psychology it must have become clear that in working out a theory of primitive sociology we had to reconstruct an important theory of human emotions, developed by one who unquestionably deserves to be ranked as one of the greatest psychologists of our time. Mr. A. F. Shand was the first to point out that in the classification of human feelings, in the construction of the laws of emotional life, we can reach tangible results only when we realize that human emotions do not float in an empty space, but are all grouped around a number of objects. Around these objects human emotions are organized into definite systems. Furthermore, Shand, in his book on *The Foundations of Character*, has laid down a number of laws which govern the organization of emotions into sentiments. He has shown that the moral problems of human character can be solved only by a study of the organization of the emotions. In our present argument, it has been possible to apply Shand's theory of sentiments to a sociological problem, and to show that a correct analysis of the change from animal to cultural responses vindicates his views to the full. The salient points which distinguish human attachments from animal instincts are the dominance of the object over the situation; the organization of emotional attitudes; the continuity of the building up of such attitudes and their crystallization into permanent adjustable systems. Our additions to Shand's theory consist only in showing how the formation of sentiments is associated with social organization and with the wielding of material culture by man.

An important point which Shand has brought out in his study of human sentiment is that the leading emotions which enter into them are not independent of each other, but that they show certain tendencies towards exclusion and repression. In the analysis which follows we shall have to elaborate on the two typical relations between mother and child on the one hand and the father and child on the other. This will also help to reveal the processes of gradual clearing off and of repression by which certain elements have to be eliminated from a sentiment as it develops.

And here we should like to add that Shand's theory of sentiments stands really in a very close relation to psycho-analysis. Both of them deal with the concrete emotional processes in the life history of the individual. Both of them have independently recognized that it is only by the study of the actual configur-ations of human feelings that we can arrive at satisfactory results. Had the founders of psycho-analysis known Shand's contribu-tion, they might have avoided a number of metaphysical pitfalls, realized that instinct is a part of human sentiments and not a metaphysical entity, and given us a much less mystical and more concrete psychology of the unconscious. On the other hand, Freud has supplemented the theory of sentiments on two capital points. He was the first to state clearly that the family was the locus of sentiment formation. He also has shown that in the formation of sentiments the process of elimination, of clearing away, is of paramount importance and that in this process the mechanism of repression is the source of conspicuous dangers. The forces of repression, assigned by psycho-analysts to the mysterious endo-psychic censor can, however, be placed by the present analysis into a more definite and concrete setting. The forces of repression are the forces of the sentiment itself. They come from the principle of consistency which every sen-timent requires in order to be useful in social behaviour. The negative emotions of hate and anger are incompatible with the submission to parental authority and the reverence and trust in cultural guidance. Sensual elements cannot enter into the rela-tion of mother and son if this relation is to remain in harmony with the natural division of functions obtaining within the household. To these questions we pass in the following chapter.

9

MOTHERHOOD AND THE TEMPTATIONS OF INCEST

The subject of the 'origins' of incest prohibitions is one of the most discussed and vexed questions of anthropology. It is associated with the problem of exogamy or of primitive forms of marriage, with hypotheses of former promiscuity and so on. There is not the slightest doubt that exogamy is correlated with the prohibition of incest, that it is merely an extension of this taboo, exactly as the institution of the clan with its classificatory terms of relationship is simply an extension of the family and its mode of kinship nomenclature. We shall not enter into this problem, especially because in this we are in agreement with such anthropologists as Westermarck and Lowie.[1]

To clear the ground it will be well to remember that biologists are in agreement on the point that there is no detrimental effect

---

[1] Compare Westermarck's *History of Human Marriage* and Lowie's *Primitive Society*. Some additional arguments will be contributed by the present writer in the forthcoming book on *Kinship*.

produced upon the species by incestuous unions.[1] Whether incest in the state of nature might be detrimental if it occurred regularly is an academic question. In the state of nature the young animals leave the parental group at maturity and mate at random with any females encountered during rut. Incest at best can be but a sporadic occurrence. In animal incest, then, there is no biological harm nor obviously is there any moral harm. Moreover, there is no reason to suppose that in animals there is any special temptation.

While with the animal then there is neither biological danger nor temptation and in consequence no instinctive barriers against incest, with man, on the contrary, we find in all societies that the strongest barrier and the most fundamental prohibition are those against incest. This we shall try to explain, not by any hypotheses about a primitive act of legislation nor by any assumptions of special aversion to sexual intercourse with inmates of the same household, but as the result of two phenomena which spring up under culture. In the first place, under the mechanisms which constitute the human family serious temptations to incest arise. In the second place, side by side with the sex temptations, specific perils come into being for the human family, due to the existence of the incestuous tendencies. On the first point, therefore, we have to agree with Freud and disagree with the well-known theory of Westermarck, who assumes innate disinclination to mate between members of the same household. In assuming, however, a temptation to incest under culture, we do not follow the psycho-analytic theory which regards the infantile attachment to the mother as essentially sexual.

This is perhaps the main thesis which Freud has attempted to establish in his three contributions to sexual theory. He tries to

[1] For a discussion of the biological nature of inbreeding, cf. Pitt-Rivers, *The Contact of Races and the Clash of Culture*, 1927.

prove that the relations between a small child and its mother, above all in the act of suckling, are essentially sexual. From this it results that the first sexual attachment of a male towards the mother is, in other words, normally an incestuous attachment. 'This fixation of libido,' to use a psycho-analytic phrase, remains throughout life, and it is the source of the constant incestuous temptations which have to be repressed and as such form one of the two components of the Œdipus complex.

This theory it is impossible to adopt. The relation between an infant and its mother is essentially different from a sexual attitude. Instincts must be defined not simply by introspective methods, not merely by analysis of the feeling tones such as pain and pleasure, but above all by their function. An instinct is a more or less definite innate mechanism by which the individual responds to a specific situation by a definite form of behaviour in satisfaction of definite organic wants. The relation of the suckling to its mother is first of all induced by the desire for nutrition. The bodily clinging of a child to its mother again satisfies its bodily wants of warmth, protection and guidance. The child is not fit to cope with the environment by its own forces alone, and as the only medium through which it can act is the maternal organism it clings instinctively to the mother. In sexual relations the aim of bodily attraction and clinging is that union which leads to impregnation. Each of these two innate tendencies—the mother-to-child behaviour and the process of mating—cover a big range of preparatory and consummatory actions which present certain similarities. The line of division, however, is clear, because one set of acts, tendencies and feelings serves to complete the infant's unripe organism, to nourish, to protect and warm it; the other set of acts subserves the union of sexual organs and the production of a new individual.

We cannot therefore accept the simple solution that the temptation of incest is due to sexual relation between the infant and mother. The sensuous pleasure which is common to both relations is a component of every successful instinctive behaviour. The pleasure index cannot serve to differentiate instincts, since it is a general character of them all. But although we have to postulate different instincts for each emotional attitude yet there is one element common to them both. It is not merely that they are endowed with the general pleasure tone of all instincts, but there is also a sensuous pleasure derived from bodily contact. The active exercise of the drive which a child feels towards its mother's organism consists in the permanent clinging to the mother's body in the fullest possible epidermic contact, above all in the contact of the child's lips with the mother's nipple. The analogy between the preparatory actions of the sexual drive and the consummatory actions of the infantile impulse are remarkable. The two are to be distinguished mainly by their function and by the essential difference between the consummatory actions in each case.

What is the result of this partial similarity? We can borrow from psycho-analysis the principle which has now become generally accepted in psychology that there are no experiences in later life which would not stir up analogous memories from infancy. Again, from Shand's theory of sentiments we know that the sentimental attitudes in human life entail a gradual organization of emotions. To these we found it necessary to add that the continuity of emotional memories and the gradual building of one attitude on the pattern of another form the main principle of sociological bonds.

If we apply this to the formation of the sexual attitude between lovers we can see that the bodily contact in sexual relations must have a very disturbing retrospective effect upon the relation between mother and son. The caresses of lovers employ not only the same medium—epidermis; not only the same situation—embraces, cuddling, the maximum of personal approach; but they entail also the same type of sensuous feelings. When therefore this new type of drive enters it must invariably awaken the memories of earlier similar experiences. But these memories are associated with a definite object which remains in the foreground of an individual's emotional interests throughout life. This object is the person of the mother. With regard to this person the erotic life introduces disturbing memories which stand in direct contradiction to the attitude of reverence, submission and cultural dependence which in the growing boy has already completely repressed the early infantile sentimental attachment. The new type of erotic sensuality and the new sexual attitude blend disturbingly with the memories of early life and threaten to break up the organized system of emotions which has been built up around the mother. This attitude, for purposes of cultural education, has become less and less sensual, more and more coloured by mental and moral dependence, by interest in practical matters, by social sentiments associated with the mother as the centre of the household. We have seen already in the previous chapters of this book how at this stage the relation between the boy and his mother is clouded over and how a reorganization of the sentiments has to take place. It is at this time that strong resistances arise in the individual's mind, that all sensuality felt towards the mother becomes repressed, and that the subconscious temptation of incest arises from the blending of early memories with new experiences.

The difference between this explanation and that of psychoanalysis consists in the fact that Freud assumes a continuous persistence from infancy of the same attitude towards the mother. In our argument we try to show that there is only a partial identity between the early and the later drives, that this identity is due essentially to the mechanism of sentiment formation; that this explains the non-existence of temptations among animals; and that the retrospective power of new sentiments in man is the cause of incestuous temptations.

We have now to ask why this temptation is really dangerous to man although it is innocuous to animals. We have seen that in man the development of emotions into organized sentiments is the very essence of social bonds and of cultural progress. As Mr. Shand has convincingly proved, such systems are subject to definite laws: they must be harmonious, i.e., emotions consistent with one another, and the sentiments so organized that they will allow of co-operation, continuity of blending. Now within the family the sentiment between mother and child begins with the early sensuous attachment which binds the two with a deep innate interest. Later on, however, this attitude has to change. The mother's function consists in education, guiding and exercising cultural influence and domestic authority. As the son grows up he has to respond to this by the attitude of submission and reverence. During childhood, that is during this extremely long period in psychological reckoning which occurs after weaning and before maturity, emotions of reverence, dependence, respect, as well as strong attachment must give the leading tone to the boy's relation to his mother. At that time also a process of emancipation, of severing all bodily contacts must proceed and become completed. The family at this stage is essentially a cultural and not a biological workshop. The father and the mother are training the child into independence and into cultural maturity; their physiological rôle is already over.

Now into such a situation the inclination towards incest would enter as a destructive element. Any approach of the mother with sensual or erotic temptations would involve the disruption of the relationship so laboriously constructed. Mating with her would have to be, as all mating must be, preceded by courtship and a type of behaviour completely incompatible with submission, independence and reverence. The mother, moreover, is not alone. She is married to another male. Any sensual temptation would not only upset completely the relation between son and mother but also, indirectly, that between son and father. Active hostile rivalry would replace the harmonious relationship which is the type of complete dependence and thorough submission to leadership. If, therefore, we agree with the psycho-analysts that incest must be a universal temptation, we see that its dangers are not merely psychological nor can they be explained by any such hypotheses as that of Freud's primeval crime. Incest must be forbidden because, if our analysis of the family and its rôle in the formation of culture be correct, incest is incompatible with the establishment of the first foundations of culture. In any type of civilization in which custom, morals, and law would allow incest, the family could not continue to exist. At maturity we would witness the breaking up of the family, hence complete social chaos and an impossibility of continuing cultural tradition. Incest would mean the upsetting of age distinctions, the mixing up of generations, the disorganization of sentiments and a violent exchange of rules at a time when the family is the most important educational medium. No society could exist under such conditions. The alternative type of culture under which incest is excluded, is the only one consistent with the existence of social organization and culture.

Our type of explanation agrees essentially with the view of Atkinson and Lang, which makes the prohibition of incest the primal law, although our argument differs from their hypothesis. We differ also from Freud in that we cannot accept incest as due to the innate behaviour of the infant. From Westermarck we differ in so far as the aversion to incest does not appear to us as the natural impulse, a simple tendency not to cohabit with persons living in the same house from infancy, but rather as a complex scheme of cultural reactions. We have been able to deduce the necessity of the incest taboo from the change in instinctive endowment which must run parallel with social organization and culture. Incest, as a normal mode of behaviour, cannot exist in humanity, because it is incompatible with family life and would disorganize its very foundations. The fundamental pattern of all social bonds, the normal relation of the child to the mother and the father, would be destroyed. From the composition of each of these sentiments the instinct of sex must be eliminated. This instinct is the most difficult to control, the least compatible with others. The temptation to incest, therefore, has been introduced by culture, by the necessity of establishing permanent organized attitudes. It is therefore, in a sense, the original sin of man. This must be atoned for in all human societies by one of the most important and universal rules. Even then the taboo of incest haunts man throughout life, as psycho-analysis has revealed to us.

10

AUTHORITY AND REPRESSION

In the previous chapter we have been mainly interested in the relation between the mother and the son; here we shall discuss that between the father and the son. In this discussion the daughter receives but little of our attention. On the one hand, as results from all that has been said above in Chap. 9, incest between father and daughter is less important, while on the other, the conflicts between the mother and the daughter are not so conspicuous. In any case what is said about mother and son and father and son can refer with little modification and on a less pronounced scale to the other set of relations. The cast of the Freudian Œdipus tragedy, therefore, in which the son again figures in relation to both parents, is anthropologically quite correct. Freud has refused even to place Electra side by side with Œdipus, and we have to countersign this act of ostracism.

In discussing previously the relation between father and son we have definitely affirmed the instinctive basis of this relationship. The human family is in need of a male, as definitely as the animal family, and in all human societies this biological need is expressed in the principle of legitimacy which demands a male as the guardian, protector, and regent of the family.

It would be useless to speculate upon the rôle of the animal father as a source of authority within the family. It seems unlikely that he should ever develop into a tyrant, because as long as he is indispensable to his children he presumably possesses a fund of natural tenderness and forbearance. When he ceases to be useful to the offspring they leave him.

Under conditions of culture the father's authority, however, is indispensable, because at the later stages, when the parents and children have to remain together for the purpose of cultural training, there is need for some authority to enforce order within the family, as indeed within any other form of human grouping. Such grouping, based on cultural and not on biological needs, lacks perfect instinctive adjustment, implies friction and difficulties and needs the legal sanction of some sort of force.

But though the father or some other male must become invested with authority at later stages, his rôle in the earlier periods is entirely different. As in the earliest stages of animal family, where the male is present to protect the pregnant and lactant female, so also in the earlier stages of the human family the father is a guard and a nurse rather than the male in authority. When he shares the taboos of pregnancy with his wife, and watches over her welfare at that time, when he becomes confined during his wife's pregnancy, when he nurses the babies, his bodily force, his moral authority, his religious prerogative and his legal power do not come into play at all. In the first place, what he has to do at those stages is regarded as a duty and not as a prerogative. In many of those intimate functions a man has to play the part of a woman—often in a somewhat undignified manner—or he has to assist her in certain tasks. Yet at the same time he is often excluded and submitted to ridiculous and humiliating attitudes—sometimes even regarded by the community as such—while his wife performs the important affairs of life. In all this, as we have repeatedly emphasized, the father acts in a meek and willing manner; he is usually very happy in performing his duties, interested in his wife's welfare, and delighted with the small infant.

The whole series of customs, ideas and social patterns imposed on the man by his culture is clearly correlated with his value to the family, with his utility to the species at that time. The father is made to behave like a loving, kind, and solicitous person, he is made to subordinate himself to his wife's organic activities, because at this stage his protection, his love, and his tender emotions make him into an efficient guardian of his wife and children. Thus here again the end of cultural behaviour among human beings is the same as that of innate endowment among animal species: this end is to shape an attitude of protective tenderness on the part of the male towards his pregnant mate and her offspring. But under conditions of culture the protective attitude has to last much longer—beyond the biological maturity of the young—while again, a much greater burden is placed upon the initial instalment of emotional tenderness. And here we find the essential difference between the animal and the human family, for while the animal family dissolves with the cessation of the biological need for parental care, the human family has to endure. After that moment the family under culture has to start on a process of education in which parental tenderness, love and care are no longer sufficient. Cultural training is not merely the gradual development of innate faculties. Besides an instruction in arts and knowledge, this training also implies the building up of sentimental attitudes, the inculcation of laws and customs, the development of morality. And all this implies one element which we have found already in the relation between child and mother, the element of taboo, repression, of negative imperatives. Education consists in the last instance in the building up of complex and artificial habit responses, of the organization of emotions into sentiments.

As we know, this building up takes place through the various manifestations of public opinion and of moral feeling, by the constant influence of the moral pressure to which the growing child is exposed. Above all it is determined by the influence of that framework of tribal life which is made up of material elements and within which the child gradually grows up, to have its impulses moulded into a number of sentiment patterns. This process, however, requires a background of effective personal authority, and here again the child comes to distinguish between the female side of social life and the male side. The women who look after him represent the nearer and more familiar influence, domestic tenderness, the help, the rest and the solace to which the child can always turn. The male aspect becomes gradually the principle of force, of distance, of pursuit of ambition and of authority. This distinction obviously develops only after the earlier period of infancy, in which, as we have seen, the father and the mother play a similar part. Later on, though the mother, side by side with the father, has to train and teach the child, she still continues the tradition of tenderness, while the father in most cases has to supply at least a minimum of authority within the family.

At a certain age, however, there comes the time at which the male child becomes detached from the family and launches into the world. In communities where there are initiation ceremonies this is done by an elaborate and special institution, in which the new order of law and morality is expounded to the novice, the existence of authority displayed, tribal conditions taught and very often hammered into the body by a system of privations and ordeals. From the sociological point of view, the initiations consist in the weaning of the boy from the domestic shelter and submitting him to tribal authority. In cultures where there is no initiation the process is gradual and diffused, but its elements are never absent. The boy is gradually allowed or encouraged to leave the house or to work himself loose from the household influences, he is instructed in tribal tradition and submitted to male authority.

But the male authority is not necessarily that of the father. In the earlier part of this book it is shown how such submission of the boy to paternal authority works and what it means. We reformulate it here in the terminology of our present argument. In societies where the authority is placed in the hands of the maternal uncle the father can remain the domestic helpmate and friend of his sons. The father to son sentiment can develop simply and directly. The early infantile attitudes gradually and continually ripen with the interests of boyhood and maturity. The father in later life plays a rôle not entirely dissimilar to that at the threshold of existence. Authority, tribal ambition, repressive elements and coercive measures are associated with another sentiment, centring round the person of the maternal uncle and building up along entirely different lines. In the light of the psychology of sentiment formation, and here I must refer to Shand's account, it is obvious that such a growth of two sentiments, each simply and internally harmonious, would be infinitely easier than the building up of the paternal relation under father-right.

Under father-right the paternal rôle is associated with two elements each of which creates considerable difficulty in the building up of the sentiment. Where this mode of reckoning of descent is associated with some pronounced form of *patria potestas* the father has to adopt the position of the filial arbiter in force and authority. He has gradually to cast off the rôle of tender and protective friend, and to adopt the position of strict judge, and hard executor of law. This change involves the incorporation within the sentiment of attitudes which are as diametrically opposed to one another as the attitude of sensuous desire and reverence within the maternal sentiment. There is no need, perhaps, to develop this point, to show how difficult it is to link up confidence with repressive powers, tenderness with authority, and friendship with rule, for on all these we have dwelt exhaustively in the earlier parts of the book. There also we have spoken of the other aspect which is always associated with father-right, even where this does not imply a definite paternal authority, for the father has always to be dispossessed and replaced by the son. Even though his powers might be limited he is yet the principal male of the older generation, represents law, tribal duties and repressive taboos. He stands for coercion, for morality, and for the limiting social forces. Here also the building up of the relationship upon the initial foundation of tenderness and effective response into an attitude of repression is not easy. All this we know.

Here, however, it is important to place this knowledge into our present argument: in the development of the human family the relation of father to offspring, instead of being based on an innate response which is closed by the departure of the mature child, has to be developed into a sentiment. The foundations of the sentiment lie in the biologically conditioned tenderness of paternal responses, but upon these foundations a relation of exacting, stern, coercive repression has to be built up. The father has to coerce, he has to represent the source of repressive forces, he becomes the lawgiver within the family and the enforcing agent of the tribal rules. *Patria potestas* converts him from a tender and loving guardian of infancy into a powerful and often dreaded autocrat. The constitution of the sentiment into which such contradictory emotions enter must therefore be difficult. And yet it is just this contradictory combination of elements which is indispensable for human culture. For the father is at the earlier stages the biologically indispensable member of the family, his function is to protect the offspring. This natural endowment of tenderness is the capital upon which the family can draw in order to keep him interested and attached to it. But here, again, culture has to make use of this emotional attitude, in imposing functions of an entirely different type upon him as the eldest male within the family. For as the children, especially the sons, grow up, education, cohesion within the family, and cooperation demand the existence of a personal authority which stands for the enforcement of order within the family and for the conformation to tribal law outside. The difficult position of the father is, as we can see, not the result merely of male jealousy, of the ill-tempers of an older man and of his sexual envy, as seems to be implied in most psycho-analytic writings; it is a deep and essential character of the human family which has to undertake two tasks: it has to carry on propagation of the species and it has to insure the continuity of culture. The paternal sentiment with its two phases, the first protective, the other coercive, is the inevitable correlate of the dual function in the human family. The essential attitudes within the Œdipus complex, the ambivalent tenderness and repulsion between son and father, are directly founded in the growth of the family from nature into culture. There is no need for an *ad hoc* hypothesis in order to explain these features. We can see them emerging from the very constitution of the human family.

There is only one way of avoiding the dangers which surround the paternal relation and this is to associate the typical elements which enter into the paternal relation with two different people. This is the configuration which we find under mother-right.

11

FATHER-RIGHT AND MOTHER-RIGHT

We are now in a position to approach the vexed problem of paternal and maternal descent, or, as it is more crisply but less precisely called, father-right and mother-right.

Once we explicitly state that the expressions 'mother-right' and 'father-right' do not imply the existence of authority or power, we can use them without danger as being more elegant than matriliny and patriliny, to which terms they are equivalent. The questions usually asked with regard to these two principles are: which of them is more 'primitive', what are the 'origins' of either, were there definite 'stages' of matriliny and patriliny?—and so on. Most theories of matriliny aimed at associating this institution with the early existence of promiscuity, the resulting uncertainty of fatherhood and thus with the need of counting kinship through females.[1] The variations on the theme *pater semper incertus* fill many volumes on primitive morality, kinship, and mother-right.

[1] See e.g. E. S. Hartland, *Primitive Society*, 1921, pp. 2, 32, and *passim*.

As often happens, the criticism which has to be directed against most theories and hypotheses must start with a definition of the concept and the formulation of the problem. Most theories imply that father-right and mother-right are mutually exclusive alternatives. Most hypotheses place one of these alternatives at the beginning, the other at a later stage of culture. Mr. S. Hartland, for instance, one of the greatest anthropological authorities on primitive sociology, speaks of 'the mother as the sole foundation of society' (*op. cit.*, p. 2) and affirms that under mother-right 'descent and therefore kinship are traced exclusively through the mother'. This conception runs throughout the work of this eminent anthropologist. In it we see mother-right as a self-contained social system, embracing and controlling all aspects of organization. The task which this writer has put before himself is to prove 'that the earliest ascertainable systematic method of deriving human kinship is through the woman only, and that patrilineal reckoning is a subsequent development' (p. 10). Remarkably enough, however, right through Mr. Hartland's work, in which he tries to prove the priority of matrilineal over patrilineal descent, we encounter invariably one statement: there is always a mixture of mother-right and father-right. In a summarizing statement indeed, Mr. Hartland says that:—'Patriarchal rule and patrilineal kinship have made perpetual inroads upon mother-right all over the world; consequently matrilineal institutions are found in almost all stages of transition to a state of society in which the father is the centre of kinship and government' (p. 34). As a matter of fact, the correct statement would be that in all parts of the world we find maternal kinship side by side with institutions of paternal authority, and we find the two modes of linking descent inextricably mingled.

The question arises whether it is at all necessary to invent any hypotheses about 'first origins' and 'successive stages' in the counting of descent and then to have to maintain that from the

lowest to the highest types of society humanity lives in a tran-
sitional state. It seems that the empirical conclusion would rather
be that motherhood and fatherhood are never found independent of each other. The logical line of inquiry indicated by the
facts would be first of all to ask the question whether there is
such a thing as matriliny independent of paternal reckoning and
whether perhaps the two types of counting descent are not
complementary to each other rather than antithetic. E. B. Tylor
and W. H. R. Rivers had already seen this line of approach and
Rivers, for instance, splits up mother-right and father-right into
three independent principles of counting: descent, inheritance
and succession. The best treatment of the subject, however, we
owe to Dr. Lowie, who has brought order into the problem and
has also introduced the very efficient terminology of *bilateral* and
unilateral kinship. The organization of the family is placed on the
bilateral principle. The organization of a clan is associated with the
unilateral kinship reckoning. Lowie[1] very clearly shows that, since
the family is a universal unit and since genealogies are universally counted equally far on both sides, it is nothing short of
preposterous to speak about the purely matrilineal or patrilineal
society. This position is entirely unassailable. Equally important
is Lowie's theory of the clan. He has shown that in a society
where in *certain respects* the one side of kinship is emphasized there
will arise groups of extended kindred corresponding to one or
other of the sib or clan organizations of mankind.

It will be well perhaps to supplement Lowie's argument and
to explain why unilateral emphasis has to be placed on the
counting of certain human relations, in what respects this is
done, and what are the mechanisms of unilateral kinship
reckoning.

We have seen that in all the matters in which the father and
the mother are vitally essential to the child, kinship has to be

[1] R. H. Lowie, *Primitive Society*, chapters on the 'Family', 'Kinship', and the 'Sib'.

counted on both sides. The very institution of the family, involv-ing always both parents, binding the child with a two-fold tie, is the starting point of bilateral kinship reckoning. If we dis-tinguish for a moment between the sociological reality of native life and the doctrines of kinship reckoning entered into by the natives, we can see that kinship is counted on both sides at the earliest stages of the individual's life. Even there, however, though both parents are relevant, their rôles are neither identical nor symmetrical. As life advances, the relation between the child and his parents changes and conditions arise which make an explicit sociological counting of kinship imperative—which, in other words, force society to frame its own doctrine of kinship. The latter stages of education, as we have seen, consist in the handing over of material possessions and of the tradition of knowledge and art associated with them. They consist also in the teaching of social attitudes, obligations and prerogatives, which are associated with succession to dignity and rank. The transmis-sion of material goods, moral values, and personal prerogatives has two sides; it is a burden on the parent who always has to teach, to exert himself, to work patiently upon the novice; it is also a surrender on the parents' side of valuables, possessions and exclusive rights. Thus, for both reasons, the lineal transmis-sion of culture from one generation to another has to be based upon a strong emotional foundation. It must take place between individuals united by strong sentiments of love and affection. As we know, society can draw upon only one source for such sentiments—the biological endowment of parental tendencies. Hence, transmission of culture in all these aspects is invariably associated with the biological relation of parent to child, it always takes place within the family. This is not enough, how-ever. There are still the possibilities of paternal transmission, maternal transmission, or else transmission in both lines. This latter can be shown to be the least satisfactory: it would intro-duce into a process which in itself is surrounded with perils, complications, and psychological dangers, an element of ambiguity and confusion. The individual would always have the choice of belonging to two groups; he could always claim possessions from two sources; he would always have two alternatives and a double status. Reciprocally, a man could always leave his position and his social identity to one of two claimants. This type of society would introduce a perpetual source of strife, of difficulty, of conflict, and as must be clear at first sight, it would create an intolerable situation. Indeed, we find our conclusion fully confirmed that in no human society are descent, succession and inheritance left undetermined. Even in such communities as those of Polynesia, where an individual can follow his maternal or paternal line alternatively, he must make his choice early in life. Thus unilateral kinship is not an accidental principle. It cannot be 'explained' as due to ideas of paternity, or to this or that feature of primitive psychology or social organization. It is the only possible way of dealing with the problems of transmission of possessions, dignities, and social privileges. As we shall see, however, this does not preclude a number of complications, supplementary phenomena and secondary reactions. There is still the choice between mother-right and father-right.

Let us have a closer look at the working of the principle of maternal and paternal kinship. As we know, the organization of emotions within the sentiment is closely correlated with the organization of society. In the formation of the maternal sentiment, as we followed it in detail in the first part of the book and as we summarized it in one of the last chapters, we are not able to see any deep disturbance by the change from the early tenderness to the exercise of authority. Under mother-right it is not the mother who wields coercive powers but her brother, and succession does not introduce any antagonisms and jealousies between the mother and her son, for here again he inherits only from her brother. At the same time the bond of personal affection and tenderness between the mother and the child is, in spite of all cultural and social influences to the contrary, stronger than between the father and the child. Nor is there any reason to deny that the obvious physical nature of motherhood may have greatly contributed towards the emphasis of the bodily identity between offspring and mother. Thus, while in the maternal tie the ideas about procreation, the tender feelings of infancy, the stronger emotional ties between mother and child would lead to a more powerful sentiment, this sentiment is in no way disturbed by the burden of legal and economic transmission which it entails. In other words, under mother-right the social decree that the son has to inherit from the mother's brother in no way spoils the relation to the mother and on the whole it expresses the fact that this relation is empirically more obvious and emotionally stronger. As we have seen in the detailed discussion of the institutions of one matrilineal society, the mother's brother, who represents stern authority, social ideals and ambitions, is very suitably kept at a distance outside the family circle.

Father-right, on the other hand, entails, as we have seen in detail in the last chapter, a definite break within the formation of the sentiment. In the patrilineal society the father has to incorporate in himself the two aspects, that of tender friend and rigid guardian of law. This creates both a disharmony within the sentiment, and social difficulties within the family by disturbing co-operation and by creating jealousies and rivalries at its very heart.

One more point may be mentioned. Even more in primitive communities than in civilized societies, kinship dominates the regulation of sexual attitudes. The extension of kinship beyond the family implies in many societies the formation of exogamy side by side with the formation of clans. Under mother-right, the prohibition of incest within the family is in a simple manner extended into the prohibition of sexual intercourse within the clan. In a matrilineal society, therefore, the building up of the general sexual attitude towards all women of the community is a continuously harmonious and simple process. In a patriarchal society, on the other hand, the rules of incest which apply to the members of the family are not simply extended to the clan but a new scheme of ideas of the sexually licit and illicit has to be built up. Patrilineal exogamy does not include the one person with whom incest should be most rigorously avoided, that is the mother. In all this we see a series of reasons why mother-right might be considered a more useful principle of social organization than father-right. The utility is obviously associated with that level of human organization where kinship plays a paramount sociological part in its narrower as well as in its classificatory form.

It is clearly important to realize that father-right also presents considerable advantages. Under mother-right there is always a double authority over the child and the family itself is cleft. There develops that complex cross-system of relationship which in primitive societies increases the strength of social texture but which in higher societies would introduce innumerable complications. As culture advances, as the institutions of clan and classificatory kinship disappear, as the organization of the local community of tribe, city, and state has to become simpler, the principle of father-right naturally becomes dominant. But this brings us out of our special line of inquiry.

To sum up, we have seen that the relative advantages of mother-right and father-right are well balanced and that it would probably be impossible to assign to either of them a general priority or a wider occurrence. The advantage of the unilateral as against the bilateral principle of kinship counting in legal, economic and social matters, however, is beyond any doubt and cavil.

The most important point is to realize that neither mother-right nor father-right can ever be an exclusive rule of counting kinship or descent. It is only in the transmission of tangible values of a material, moral or social nature that one of the two principles becomes legally emphasized. As I have tried to show on other occasions,[1] such a legal emphasis brings with it certain customary traditional reactions which tend to a certain extent to obliterate its one-sided working.

Returning once more to our starting-point, that of the criticism expressed by Dr. Jones on the conclusions reached in the previous parts of the book, it can now be seen that the appearance of mother-right is not a mysterious phenomenon brought about by 'unknown social and economic reasons'. Mother-right is one of the two alternatives of counting kinship, both of which shows certain advantages. Those of mother-right are perhaps on the whole greater than those of father-right. And among them unquestionably we have to mention the central point which has been brought out in this chapter: the value which it has in eliminating the strong repressions in the paternal sentiment and in placing the mother in a more consistent and better adapted position within the scheme of sexual prohibitions in the community.

[1] *Crime and Custom in Savage Society*, 1926; *Nature*, supplement of 6th February, 1926; and article of 15th August, 1925.

12

CULTURE AND THE 'COMPLEX'

We have now covered the field of our subject: the change in instinctive endowment correlated with the transition from nature to culture. We can briefly indicate the course of our argument and summarize our results. We started with psycho-analytic views on the origins and history of the complex. In this we came upon a number of obscurities and inconsistencies. The concept of the *repression of already repressed* elements; the theory that ignorance and matriliny were *devised* as means of deflecting hatred; the idea that father-right is a happy solution of most difficulties in the family; were all difficult to reconcile with the general doctrine of psycho-analysis as well as with fundamental anthropological facts and principles. It was found also that all these inconsistencies result from the view that the Œdipus complex is the primal cause of culture, that it is something which preceded and produced most human institutions, ideas, and beliefs. In attempting to find in what concrete form this primordial Œdipus complex has originated according to psycho-analytic theory, we came upon Freud's hypothesis of the 'primeval crime'. Freud regards culture as a spontaneously generated reaction to the crime and he assumes that the memory of the crime, the repentance and the ambivalent attitude have survived in a 'Collective Unconscious'.

Our utter incapacity to accept this hypothesis forced us to examine it more closely. We found that the totemic crime must be imagined as a dividing event between nature and culture; as the moment of cultural beginning. Without this assumption the hypothesis has no meaning. With it the hypothesis falls to pieces because of the inconsistencies involved. Having found that in Freud's hypothesis as in all other speculations on the early form of the family, the capital mistake is made of ignoring the difference between instinct and habit, between the biologically defined reaction and the cultural adjustment, it became our task to study the transformation of family ties due to the passage from nature to culture.

We attempted to ascertain the essential modification in innate endowment and to show what were the consequences of it to human mentality. In the course of this we naturally came upon the most important psycho-analytic problems, and we were able to offer a theory of the natural formation of the family complex. We found the complex as an inevitable by-product of culture, which arises as the family develops from a group bound by instincts into one which is connected by cultural ties. Psychologically speaking, this change means that a cohesion by a chain of linked drives is transformed into a system of organized sentiments. The building up of sentiments obeys a number of psychological laws which guide the mental ripening so as to eliminate a number of attitudes, adjustments, and instincts from a given sentiment. The mechanism of it we found in the influence of the social environment, working through the cultural framework and through direct personal contacts.

The process of elimination of certain attitudes and impulses from the relation between father and child and mother and child present a considerable range of possibilities. The systematic organization of impulses and emotions may be carried out by a gradual drawing off and waning from certain attitudes, by dramatic shocks, by organized ideals, as in the ceremonies, by ridicule, and public opinion. By such mechanisms we find, for instance, that sensuality is gradually eliminated from the child's relation to its mother, while often tenderness between father and child is replaced by a stern and coercive relation. The way in which these mechanisms operate does not lead to exactly the same results. And many maladjustments within the mind and in society can be traced back to the faulty cultural mechanism by which sexuality is suppressed and regulated or by which authority is imposed. This we have presented with great detail in a small number of concrete cases in the first two parts of the book. This again has been theoretically justified in this last part.

Thus the building up of the sentiments, the conflicts and maladjustments which this implies, depend largely upon the sociological mechanism which works in a given society. The main aspects of this mechanism are the regulation of infantile sexuality, the incest taboos, exogamy, the apportionment of authority and the type of household organization. In this perhaps lies the main contribution of the present memoir. We have been able to indicate the relation between biological, psychological and sociological factors. We have developed a theory of the plasticity of instincts under culture and of the transformation of instinctive response into cultural adjustment. On its psychological side our theory suggests a line of approach which, while giving full due to the influence of social factors, does away with the hypotheses of 'group mind', the 'collective unconscious', 'gregarious instinct', and similar metaphysical conceptions.

In all this we are constantly dealing with the central problems of psycho-analysis, the problems of incest, of paternal authority, of the sexual taboo and of the ripening of the instinct. In fact the results of my argument confirm the general teachings of psycho-analysts on several points, though they imply the need of serious revision on others. Even on the concrete question of the influence of mother-right and its function, the results which I have published previously and the conclusions of this book are not entirely subversive of psycho-analytic doctrine. Mother-right, as has been remarked, possesses an additional advantage over father-right in that it 'splits the Œdipus complex', dividing the authority between two males, while on the other hand it introduces a consistent scheme of incest prohibition in which exogamy follows directly from the sexual taboo within the household. We had to recognize, however, that mother-right is not altogether dependent upon the complex, that it is a wider phenomenon determined by a variety of causes. These I have tried to state concretely in order to meet Dr. Jones's objection that I assume this appearance for unknown sociological and economic reasons. I have tried to show that mother-right can be made intelligible as the more useful of the two alternative forms of reckoning kinship. The real point, as we saw, is that the uni-lateral mode of counting relationship is adopted in almost all cultures but that among peoples of low cultural level the maternal line shows distinct advantage over the paternal one. Among these signal characteristic advantages of mother-right we find its power to modify and split the 'complex'.

I should add that from the point of view of psycho-analytic theory it is difficult to explain why the complex as such should be harmful. After all, to a psycho-analyst, the Œdipus complex is the *fons et origo* of culture, the beginning of religion, law, and morality. Why should there be any need to remove it? Why should humanity or the 'collective mind' have 'devised' any means to break it up? To us, however, the complex is not a cause but a by-product, not a creative principle but a maladjust-ment. This maladjustment assumes a less harmful form under mother-right than under father-right.

These conclusions were first set forth in two articles which appeared separately a few years ago and are now reprinted as Parts I and II of this volume. Here again in dealing with the general problem, we have found certain confirmations of psycho-analytic theory, if this be taken as an inspiration and a working hypothesis and not as a system of dogmatic tenets.

Scientific research consists in collaboration, in a give and take between various specialists. The anthropologist has received some help from the psycho-analytic school and it would be a great pity if the exponents of this latter refused to collaborate, to accept what is offered in good faith from a field where, after all, they cannot be at home. The advancement of science is never a matter of simple progress in a direct line. In the conquest of a new domain, claims are often pegged out on which the barren soil will never yield a return. It is as important for a student or for a school to be able to withdraw from an untenable position as to pioneer ahead into new fields of discovery. And, after all, it should ever be remembered that in scientific prospecting the few grains of golden truth can only be won by the patient washing out and rejection of a mass of useless pebbles and sand.

INDEX

Kula: dreams 78; magic 103
Kwoygapani: magic 91

Lang, Andrew 125n., 197
language 144, 145, 151
legitimacy, postulate of 169–70
'libido' 1, 28, 193
Lowie, R. H. 125n., 147n., 191n., 208
luguta (sister) 88

magic: 78, 88, 94–6; black 72–3, 83; love 70, 78, 98–102, 103, 105–6; *suvasova* 82; shipwreck 98; — and myth 94–105; magical filiation 103–4; *see also Kwoygapani, waygigi*
Mailu: neurasthenics in 74
Malasi: clan, and incest 81
Marett, R. R. 125n.
marriage: in Trobriands 9–11, 37; animal 160–1, 162; human 161–4, 178–80, 182–3
'mass psyche' 124–6
material culture 144–5, 152
maternal instinct 11, 18, 19, 128, 165–7; *see also* affection
mating 156–7, 178–9, 180, 182
matriliny 5, 8–9, 60, 84; — and myth 88–94; *see also* mother-right
matrimonial response 161, 168
McDougall, W. 139
medium (spiritualistic) 73, 81
miracle 104–5
missions: and native morality 75
Mokadayu, story of 81

Moll, A. 28n., 40
morals 145, 201–2
Morgan, L. H. 176
mother: and son 30–1, 32, 50, 63, 175, 193–8, 210–11; *see also* incest; — and daughter 30–1, 52
motherhood: in savage and civilized society compared 11–19, 23–4
mother-right: 84; 'origin' 110–12, 116–17, 136, 217; and father-right Part IV, Ch. 11; *see also* matriliny
mother's brother 9–10, 12, 37–9, 55, 63, 83, 91–2, 94, 97, 203; *see also* Tudava
myth 6, 88–106; classification of 88; interpretation of 93; — of flying canoe 95–8; — of salvage magic 98; — of love magic 100–2; and ritual 105; *see also* fire, magic, Tudava

nagowa (mental disorder) 73
nakedness: no taboo in Melanesia 44
neurosis: among Melanesians 71–5; interpretation of 135
'nuclear complex' 5, 60, 111, 137–40; varies with social strata 12–13; with constitution of family 64–5, 114; Jones's view of 111; *see also* complex
nursing of child 18–19, 20, 166, 193–4

obscenity 85–8

Œdipus complex: 3, 6, 7, 52–3, 62, 64, 69, 109, 145, 193, 199, 205, 214, 217; product of a patriarchal society 6, 60, 133–6, 137; assimilation of 133–4; assumed universality 111–13, 114–17, 126, 130, 135–6

parental love: among animals 128, 165–6; — in man 166
parricide 84; primeval 121, Part III, Ch. 4 and Ch. 5; critique of 136, 215
paternity: biological foundation of 165, 168–71; cultural reinforcement of 171–2; *see also* fatherhood
patria postestas *see* father, authority
patriarchy 133–5
peasant family 12–13, 14, 25, 29, 31, 35, 37, 41–3, 51–2
perversions, rare in Trobriands 46, 74–5
physiological fatherhood *see* fatherhood
Pitt-Rivers, G. 19n., 147n., 192n.
plasticity of instincts 159, 164, 171–2, 177, 186, 216, Part IV, Ch. 7; *see also* instinct
Ploss-Renz 28n.
pokala (payment) 97
pregnancy 17–18, 162, 166, 168–9, 170; *see also* taboo
primal horde *see* family, Cyclopean
property, in magic 96–7, 102
psycho-analysis: relation to biology and sociology 1, 109; to

anthropology 6–7, 93–4, 110, 113, 218; — and family 4, 60, 64; — and myth 106; — and theory of A. F. Shand 190
puberty 48–56; of civilized boy 49–51; of civilized girl 52–3; in Trobriands 53–8

Rank, Dr O. 6, 19
repression 32–3, 57, 64, 189–90, 213; — and neurosis 74; — and dreams 76–7; — and abuse 85–8; — and myth 90–1; — of knowledge of paternity 110–11, 116–17; — and the complex 115–17, 136, 137–9
Rivers, W. H. R. 208
Robertson Smith, W. 119
rut 154, 157–8, 161

selective mating *see* mating
sentiment 4, 60, 139–40, 152, 163, 189–90, 195–6, 204–5, Part IV, Ch. 8; *see also* instinct, complex
sex confidences 30n.
sex latency period 40–4, 46, 62
sexuality of children 9, 28–30, 40–1, 44–7, 61, 72, 216; *see also* games
sexual desire: in marriage 184
sexual dreams 79–80
sexual impulse: 154–8; control of 158–9
sexual relations: 26, 156, 160–4, 182–5; and mother 194–5
sexual rivalry 30–2